ARBITRATION AND RENT REVIEW

ARBITRATION AND RENT REVIEW

Third Edition

Ben Beaumont

B.Sc., FRICS, FAMINZ, WIPO and CIETAC Panellist, Barrister
Official Delegate to the UNCITRAL 2003 Forum
Fellow of Chartered Institute of Arbitrators and Council Member 1981–99
Member of Lamb Chambers, Temple

Foreword
by
Lord Bingham of Cornhill

2004

Routledge
Taylor & Francis Group

LONDON AND NEW YORK

First published 1990 by Estates Gazette

Second edition 1993

Third edition 2004

Published 2014 by Routledge

2 Park Square, Milton Park, Abingdon, Oxon OX14 4RN

711 Third Avenue, New York, NY 10017, USA

Routledge is an imprint of the Taylor & Francis Group, an informa business

ISBN 978-0-728-20425-6 (pbk)

*

Dedicated

to

Andrea, Caroline, Daniel, Earl, and Kelly

*

AUTHOR'S ACKNOWLEDGEMENTS

I apologise for the space of ten years between this edition and the last. It will not occur again. I completed these revisions and additions before the retirement of Colin Greasby of the Estates Gazette, to his surprise and my pleasure. I have now included cases from Australia, New Zealand, Canada, Hong Kong, and Scotland, as well as England and Wales, up to the spring of 2003.

On this occasion I have managed to become computer manipulative, so all errors are mine. However, guidance upon the structure and all general issues has been the superlative work of Sue Godfree of Leaf Coppin.

Finally, I am very proud to be able to thank Lord Bingham of Cornhill, a former President of the Chartered Institute of Arbitrators, for agreeing to write a Foreword to this edition.

FOREWORD

Rent review clauses are nowadays a very standard feature of commercial leases. The reason is not far to seek. Given the vagaries of inflation and changing market conditions, a rent which reflects the going rate when agreed at the beginning of a term may, within quite a short time, cease to do so, probably because it is too low. Then, when the time for review comes, the landlord is likely, in the nature of things, to ask rather too much, the tenant to offer rather too little. So there is need for a procedure to revise the contractual rent at the agreed intervals, to bring it back into line with the going rate.

In some cases such reviews may involve very large figures indeed. Then the landlord and the tenant may well spare no expense in retaining the most skilful, experienced, and knowledgeable counsel, solicitors, surveyors, and arbitrators or independent experts. But there are many much smaller cases, not warranting such lavish expenditure, but none the less involving sums which may have a crucial bearing on the business interests, or even the survival, of landlord or tenant. It is by those engaged in rent reviews of this kind that the Third Edition of this book will be particularly welcomed.

For in this book the author explains the process of rent review arbitration or expert determination in an ordered, simple and readily intelligible way, while not ignoring the undoubted complexities of the subject. He points out the pitfalls. He draws attention to the leading legal authorities, while keeping his text commendably free of unnecessary legal jargon. He gives the reader the benefit of his own opinion, as an experienced arbitrator, on points of doubt and difficulty.

It is in the interest of landlords and tenants, and of society as a whole, that contested rent reviews should be carried out fairly, professionally, inexpensively and without recourse to the courts. I am sure that this new edition, like its predecessors, will contribute to that end.

Tom Bingham
December 2003

CONTENTS

TABLE OF CASES

INTRODUCTION

In the 1950s and 1960s there was normally no review of the rent during the period of a lease. This method of receiving a return on capital invested was excellent whilst there was no regular reduction in the value of the pound. As soon as inflation became persistent, businessmen realised that a fixed return on their investment meant that in real terms they were receiving an ever-smaller return. In order to meet this problem, a procedure for the review of the rent payable was inserted in the lease. Although a procedure for rent reviews is not a new phenomenon, it generally comes into force only in times of growth in the rate of inflation.

At present, the trend is that reviews of rent should take place every four or five years. Unfortunately for the tenant, reviews of rent take into account not only the identifiable factor of diminution in value brought about by the devaluation of the pound, but also the indefinable factor that may cause a certain location to increase in popularity. This will have an effect upon rental values.

For an excellent introduction as to the purpose of the rent review see *British Gas Corporation v Universities Superannuation Scheme Ltd.* [1986] 1 EGLR 120. Further guidance is given in *British Airways plc v Heathrow Airport and Another* [1992] 1 EGLR 141 from which are taken the following abbreviated principles. These appear to be a useful template for lawyers, arbitrators, and experts.

1. It is necessary to determine the intention of the parties at the time of entering into the agreement as far as can be deduced from the particular clause. The arbitrator must first examine the clause itself and then the surrounding circumstances.
2. The rationale for the rent review clause is to ensure that the rent payable reflects the changes in the value of money and of the demised premises during the period of the lease.
3. The role of the court should be limited to giving guidance as to how to interpret the clause. No directions should be given to the expert or arbitrator as to how to carry out his function.
4. It is only the real circumstances affecting the property which should be taken into account. But first there must be taken into consideration the assumptions and disregards expressly and impliedly set out in the rent review clause.
5. Usually the property valued will be the actual existing premises. However the requirements of the assumptions and disregards may require a valuation which departs from reality.

6. It may be necessary to disregard the actual state or condition of the premises. Thus a former warehouse converted into luxury office space may be required to be valued as if the former warehouse were still the reality.

How is the Rent Reviewed?

Review Clause in Lease

In the lease there will usually be a clause which indicates how the rent will be reviewed. If by some chance there is no clause to this effect in the lease in question, then it is very doubtful indeed whether any court would in fact imply such a clause.

As to the construction of a rent review clause, apposite comments can be found in *Co-operative Wholesale Society Ltd. v National Westminster Bank plc* [1995] 1 EGLR 97: 'There are no special rules for the construction of rent review clauses.' However, the court cited *British Gas Corporation* (see above):

> ... Of course, the lease may be expressed in words so clear that there is no room for giving effect to such underlying purpose. Again, there may be special surrounding circumstances which indicate that the parties did intend to reach such an unusual bargain. But in the absence of such clear words or surrounding circumstances, in my judgment the lease should be construed so as to give effect to the basic purpose of the rent review clause and not to confer on the landlord a windfall benefit which he could never obtain on the market if he were actually letting the premises at the review date ...

Further guidance was given in *Basingstoke and Deane Borough Council v Host Group Ltd.* (see p. 70).

Implementation of the Review Procedure and 'Time of the Essence'

Where there is a review clause within the lease, the landlord notes the procedure set out in the clause and ensures that either his staff or his advisers are informed as to any timetable which is set out in the clause. A typical clause might indicate that the landlord shall notify the tenant of any increase in rent in the period six months before the new rent becomes effective. In the author's view, it is essential that timetables are treated as critical until it is mutually agreed otherwise.

Valuation

At the relevant time, the landlord will, either from his own knowledge or having taken advice from his management surveyors, determine the value of the new rent, having taken into account the various assumptions and disregards.

Notice ('Trigger Notice')

As soon as the new rental value is received, the landlord, or his agent, must write formally and clearly to the tenant stating the rent required from the date from which the reviewed rent is payable. This is a notice or 'trigger notice'.

Privilege   Privilege means protecting a document from being produced in a court of law or arbitration by certain legal rules. It is suggested that landlords require their solicitors to instruct surveyors to prepare the initial outline valuation on the specific basis that a dispute will be forthcoming. If this is done the actual valuation may be privileged, though this has not been clarified by the courts.

'Without prejudice' and 'subject to contract'   The trigger notice should not be headed 'subject to contract' or 'without prejudice' as this may give the impression that such a notice is merely an opening gambit and not a formal notice.

Counter-Notices

A tenant will receive the notice. The first step for the tenant is to examine the clause in the lease under which the procedure is implemented and to ensure that a formal reply is made to the landlord's notice in good and effective time. All the tenant has to do to protect himself is to reply formally that he has received the notice of the landlord and objects to the rent stated therein, and seeks the matter to be resolved by an independent expert or arbitrator.

Failure to Serve Counter-Notice

If the tenant does not take this simple and rudimentary step of objecting, it can and does occur that the initial rent proposed by the landlord will become the effective rent payable. This can have a quite disastrous financial effect upon the tenant. For example, take premises let at an initial rent of £10,000 per annum exclusive. At the first review after four years from the commencement of the lease, the landlord submits a reviewed rent at £20,000 p.a. exclusive. The actual rental value may not be more than £14,000 p.a., but the tenant may ignore the time limits set out in the lease until too late and find himself required to pay £6,000 more than the true rental value merely by failing to act.

Attempt to Agree Rent

After the exchange of correspondence has taken place, the parties, if they have not already done so, appoint advisers to act on their behalf; they are instructed to attempt to agree a new rent. Time limits set out in the lease as to what should occur in the event of failure to agree within a certain time must be adhered to. Such adherence to strict formality will not prevent agreement of the revised rent by amicable means, but will ensure that neither party is prejudiced by failure to adhere to time limits.

That agreement, if reached, should be formally set down and attached to the lease document, signed and witnessed by both parties or their agents.

Requirement to Appoint an Independent Expert or an Arbitrator

Where the parties fail to agree, as will be the case on some occasions, then the appointment of an independent expert or arbitrator to decide the level of the revised rent must be made.

The lease may state which of these is to be appointed; in the event of the lease not being clear (much to be regretted) the parties themselves should attempt to agree the procedure without going to court.

Nomination by parties   It is quite usual for the parties themselves to agree upon the name of the arbitrator or expert involved. The normal procedure is for the landlord to submit a list of names, and for the tenant to agree upon one of the names submitted to be the arbitrator or expert in question.

Nomination by others   In the event of the parties not agreeing on a name, the alternative procedure as may be set down in the lease for the appointment of an arbitrator or independent expert by the president of the Royal Institution of Chartered Surveyors, or the president of the Chartered Institute of Arbitrators, will be required. Where there is uncertainty as to the machinery relevant for the appointment of an arbitrator within the lease and the court has decided as to the method of appointment, it is more likely that the actual nomination will be referred back to one of the professional bodies than remain in the hands of the court itself.

Differences between a Decision by an Arbitrator and One by an Independent Expert

Criteria for decision by an arbitrator   It is generally agreed that an arbitrator must act in a judicial capacity, is required to receive evidence from both parties if offered and to take note of that evidence, and subsequently must issue an award which is final and binding. He is also required to give reasons for his award if requested so to do by either or both of the parties involved.

Criteria for decision by an independent expert   An independent expert on the other hand is not required to act judicially, is not required to receive evidence from the parties, nor is he required to take note of any evidence if he does receive it from the parties. He is, however, required to make a formal and binding award. There is no obligation upon the independent expert to give reasons for his award, although the parties may in their initial appointment of the independent expert require him to give reasons.

Reasoned awards   Although there is a tendency among independent experts to give reasoned awards, there is still considerable unwillingness to do so because it is quite clear that where reasons are given for such awards, and those reasons are found to be faulty, then the expert may be sued for negligence, whereas an arbitrator cannot be sued for negligence.

Decision by an Independent Expert

The procedure for a DETERMINATION* of rent by an independent expert may not differ in the initial stages from that by an arbitrator. The expert may well call the parties, or their representatives, to attend upon him. At that meeting the expert will either by consent, or by order, issue what is known as an ORDER FOR DIRECTIONS. This Order for Directions will set out the timetable for the determination, i.e. the dates by which the written submissions (should they apply) will be made and exchanged, subsequently the dates by which comments upon those submissions shall be forthcoming, and when both comments shall be exchanged.

Form of hearing The majority of rent review determinations, whether decisions by independent experts or arbitrators, are not carried out in the form of a hearing with parties giving oral EVIDENCE on either side. Evidence is submitted in the form of WRITTEN REPRESENTATIONS. Thus the landlord will submit a written report which will state the evidence and comparable evidence (see below, p. 62) upon which he wishes to rely to prove the rental value. He will send this report to the tenant with a copy to the arbitrator or expert. The tenant will reply with a similar report following the same procedure. Subsequently, if the parties have agreed, both landlord and tenant will exchange further reports commenting upon each other's evidence. Thereafter the procedure differs.

Determination of value If an arbitrator is determining the rental value then that value must be determined upon the evidence set out in the written submissions and upon no other evidence whatsoever, whereas if the independent expert is determining the rent the expert may entirely ignore the written submissions and determine the rent by reliance upon his own expertise. This expertise may result in a rental level which is either higher or lower than the levels submitted by either party.

Post determination Once the expert has decided upon the level of the rent, he informs the parties that he has made his decision and that his award is awaiting collection. One or both of the parties collects the award upon payment to the expert of his fees.

Differences of control The procedure with the use of an independent expert is simple and straightforward. However, a major problem is that there is no judicial control of this procedure. It is a point of importance because there is no recourse to justice other than for a negligent act.

*Words printed in small capitals are to be found in the Glossary at p. 123.

Arbitration

Written representations   An arbitration may be decided either by written representations or by means of an oral hearing. In the case of written representations, the arbitrator will, once he has been appointed, also issue an Order for Directions setting out the timetable for the receipt and exchange of the various reports. When this timetable has been completed, the arbitrator will issue his award in a very similar way to that of the independent expert.

Oral hearings or not   Where the parties and/or the arbitrator decide that the determination should be made by means of an oral hearing the procedure is slightly more formal. There has been discussion as to whether or not oral hearings should take place. The author supports the view that whether complicated issues of law exist or not an oral hearing does give the opportunity to both sides to cross-examine the experts of the other party properly as to the evidence that they are putting forward. It may well be only under cross-examination that essential valuation factors emerge which explain discrepancies in rental levels not apparent within the body of the report. Indeed the fact that there is a POINT OF LAW in dispute does not of itself necessitate an oral hearing. There is a body of legal opinion which supports the view that closing speeches by counsel which will of necessity encompass points of law should be submitted in a written form. In some courts this procedure is already being implemented.

Preliminary meeting with an arbitrator   Where an oral hearing is to take place the arbitrator will normally call what is known as a preliminary meeting at which the parties and/or their representatives will be present. At this meeting procedural matters will be discussed and a timetable for the exchange of various documents will be set down in an Order for Directions, and a date for the hearing agreed.

Location of hearing   Most arbitrators will look to the party seeking to initiate arbitration proceedings to supply a suitable room at a place convenient to all parties (including the arbitrator). A small boardroom in the offices of one of the parties should be sufficient and there should also be available if possible two additional small rooms, one in which the arbitrator may retire should he be required so to do, and one for the other party to the arbitration to use for conferences. In the event of such a room not being available the use of a nearby hotel will generally be sufficient.

Exchange of correspondence   When the surveyors to both sides have collated their evidence and prepared a report, that report should be sent to their clients, their clients' solicitor, and counsel if employed. The exchange of these reports should not take place until authorised by the client and also counsel where utilised. It is crucial that when writing to the arbitrators or to the other side, copies of such correspondence are always sent to the third party. Arbitrators have had their awards put aside and the whole procedure has had to start again where it has subsequently been found that they have been receiving from one side letters of which copies have not been passed to the other side. This problem is circumvented from the beginning if both sides automatically send copies of correspondence to all parties.

Continuing negotiations ('without prejudice')   Continuing correspondence, in the hopes of settling the matter before the hearing, should always be headed 'without prejudice' and contain clear and concise attempts to agree the revised rent. These items of correspondence will not be brought to the attention of the arbitrator as they are protected (privileged).

If they reach the arbitrator, either as the result of an accidental enclosure or disclosure, or by a deliberate attempt to destroy the credibility of the arbitrator, the arbitrator has a clear course of action. He must draw attention to the receipt of the information in writing. At the same time once he has the slightest idea that he is in possession of such information he must return it the sender and ask for an explanation. Then he should make abundantly clear that he can and will put from his mind the glimpses that made him aware of the content. Judges are constantly required to do the same so why should arbitrators be placed in a more vulnerable situation?

Exchange of reports and agreement of facts as facts   It should be stressed that when their reports have been exchanged, the landlord and tenant should attempt to agree as much as they may between them as to the various items of comparable evidence, e.g., the size of the building, the form of lease and, where there is a PREMIUM, for what precisely that premium was paid.

Procedure at hearing   At the hearing the landlord or his representatives will give evidence on those areas of the report which have not been agreed between the parties. When the representative of the claimant has given evidence, the tenant or his representative will cross-examine the landlord's witness, and, when that cross-examination has been completed, the landlord will be allowed a re-examination of his witness on any points arising from the cross-examination. When the landlord's witnesses have given their evidence then it is the turn of the tenant. When both sides have given evidence the tenant will sum up his evidence, and finally the landlord will produce the closing speech on his own behalf.

Publication of award   Some time later (unless there is some matter upon which an interim award is required) the arbitrator will indicate to the parties that his award is ready to be taken up. Once it has been taken up, the rent is generally effective from the quarter day after the date of publishing the award. Publication takes place on the date that the arbitrator notifies the parties that the award is ready for collection.

Costs

These are the costs of both the preparation for the hearing and the hearing itself, including the arbitrator. These costs will be relevant whether the valuation is decided by an arbitrator or an independent expert.

Some leases will stipulate that in any event each side will pay their own costs. This stipulation is binding where there is a REFERENCE to an independent expert. When there is a reference to arbitration such a clause is void. The reason for this is that the discretion of the arbitrator to award costs as he thinks fit must not be fettered before the reference has started. However, it is usual that the parties agree to an equal division of costs once the arbitrator has been appointed. This latter course of action is permitted.

The costs involved should be estimated before a reference is made. It may well be that in a dispute over a valuation of under £15,000 per annum, the costs attaching to a hearing and the requirements thereof are greater than the resulting rental benefit. This will not often be the case in the present property market.

One point cannot be made too strongly: it is considered more effective to obtain one fully authenticated and supportive comparable than a multitude of irrelevant comparables, however tempting such comparables may be. Large savings in costs will also result. It is now clear from recent judgments that arbitrators are making great efforts to issue directions which make clear in what form the comparable should be and how it should be presented. Time and again these directions are ignored. It is not unreasonable for an arbitrator to penalise a party in costs where that party has refused to comply with an order for directions. It also might be asked whether parties should refuse to pay the costs of an arbitrator where the arbitrator has failed to adhere to his own directions.

Something must be done about costs in arbitration. It is being marketed as cheap when compared with litigation. A rent review carried out recently had a claim for just £20,000. The arbitration was to be determined upon documents only. No submissions were made. There was no reason for their absence. Simple pleadings and five affidavits exhibiting invoices were prepared. The claimant's costs were £41,000, for a major legal practice. Those of the arbitrator were £29,000 and those of the in-house staff of the respondent, assisted by a quantity surveyor, were £22,000.

Interest

As to whether and when interest is payable on the outstanding difference between the existing rental level and the new revised rental level, careful examination of the relevant clause within the lease will (or may) resolve this matter. There are unfortunately leases still being issued which make no mention as to whether interest is payable on any such differences; or if they do state the procedure, they do not take into account any greater time than the period from the date of publication to the next quarter day. This ignores the frequent occurrence that arbitration may not take place for up to a year or more after the date when the reviewed rent is actually payable. This is another reason why it is essential for landlords to adhere to the timetables set down within the lease.

Reasoned Awards

If an award has been published without reasons, a non-speaking award, that will usually be the end of the matter and the tenant will commence paying the reviewed rent. If, however, reasons have been requested by one or more of the parties and a reasoned or SPEAKING AWARD has been given, then those reasons will be examined by the parties. One reason for seeking such an award is that it may well reveal areas in the decision where the arbitrator may have made errors on matters of law or, indeed, findings of fact which no reasonable arbitrator could have possibly made on the evidence submitted.

Appeals to Court

Appeals from the award of an arbitrator are possible. The grounds for such appeal are limited. The Arbitration Act 1996, together with amplification by subsequent case law, has set strict criteria.

An appeal may exist from a reasoned award on a point of law. That point of law must be on a matter of general interest to others, not merely to the particular parties – also the rights of the parties must be substantially affected by the outcome. Another ground for appeal is where the finding is obviously wrong. This is a highly debatable area.

It is normally impossible to mount an appeal on an unreasoned award unless there has been misconduct by the arbitrator.

Misconduct

Misconduct does not mean here behaviour of a disgraceful nature. Generally it means that in some way the arbitrator has failed to follow the rules of natural justice, or it may be a simple procedural issue, hence the possibility of appeal, and, depending on the view of the court, the award will be either set aside or remitted.

Areas of misconduct range from taking into account the incorrect retail price index; failing to send copies of correspondence from one side to the other; failing to inspect all, or some, of the premises; giving evidence to oneself, as arbitrator, as to the rental levels, and relying on that evidence, without inviting the comments of the parties; ignoring the directions issued to the parties; taking into account and relying fundamentally upon opinion as to value given by non-experts; and failure to determine all matters in dispute. This last item can be rectified within one month either by the arbitrator himself or at the request of a party.

It may be seen that most of the areas of misconduct are easily rectified by utilisation of a bit of care and forethought or even, dare it be said, by proper business management. It must be emphasised that cases where misconduct occurs represent a minute proportion of all arbitrations.

Remission or Setting Aside of Award

REMISSION OF AWARD means that the award is returned to the arbitrator for a fresh award to be issued by him taking into account the altered factor indicated by the court.

SETTING ASIDE AN AWARD means that the whole arbitration leading to the award is cast in doubt and that a new arbitration must take place. This will generally be under the guardianship of a fresh arbitrator.

Certification of Rent

Finally, when the dust has settled and the new revised rent is accepted by the parties, that new figure should be certified in writing and attached to the lease. Where any points of law have been decided then the award which encompasses those points of law should also be attached to the lease. This must be done not least because it could well be that both the landlord and/or the tenant are different parties by the time the next review comes round.

Relevance to Future Rents

Although a decision by an arbitrator is not binding on the parties at the next review, it is a strong indication of how one arbitrator views the disputed area within the lease. Where the award indicates that the arbitrator has heard clear legal representation from both sides it may well be that fresh parties to the scene may not wish to re-argue the matter but may be content to rely on that previous award as to the decision on the area in dispute or doubt.

The book is structured by taking each legal aspect of a rent review and arbitration in, as it were, chronological order and discussing its meaning and application through examination of actual court decisions.

The various cases that are used to illustrate points made in the text are not exhaustive and although some are contemporary, as can be seen from the dates, others have been chosen because they are of general importance or specific relevance.

REQUIREMENTS OF AN ARBITRATION AGREEMENT

Excomm Ltd. v Bamaodah [1985] 1 Ll R 403
Only one fact is of interest in this context, namely, that a standard form of contract stated that all disputes under the contract were to be referred to arbitration. This form was not signed.

The Court of Appeal decided that an arbitration agreement existed between the parties even though the contract was not signed.

An agreement has now been established. It may be an oral agreement, in which case the parties will have to agree as to what were the terms of agreement. The courts in England and Wales are very unwilling to enforce an oral agreement unless there is almost no dispute as to its existence and the terms thereof In New Zealand, however, the Arbitration Act 1996 has been adapted from the standard UNCITRAL (United Nations Commission on International Trade) Model Arbitration Law format to include a REFERENCE to oral contracts.

However where the agreement is in writing, which is the normal state of affairs, then it will be possible to examine the document in which the terms of agreement are set out in order to decide the circumstances as to the existence, or otherwise, of the crucial machinery for the implementation of the rent review.

RENT REVIEW CLAUSE LACKING IN FORMULA AND MACHINERY

Thomas Bates & Son Ltd. v Wyndham's (Lingerie) Ltd. [1981] 1 WLR 505
In a lease where there was a reference to the fact of a review and the fact that subsequently such rents relevant to the review should be such as 'shall have been agreed', the question arose as to whether the courts are able to imply machinery and a formula. It was decided that machinery and a formula could be implied because it was clear that a review was intended. However, as will be noted, even when there are what might be regarded as clear indications that the rent will be reviewed the courts will not always agree.

OMISSION OF RENT REVIEW MACHINERY – RECTIFICATION NOT ALLOWED

Kemp v Neptune Concrete Ltd. [1988] 2 EGLR 87
An initial term of six years was extended to twelve years. Provision was made for a rent review after the first three year period. Thus, there was a provision in

the initial term for further reviews. However, in spite of the extended period, no provision was made for such reviews.

No reference was made to the provision in the correspondence between the parties' solicitors. There were indirect references to rent reviews in the plural as opposed to the singular. The landlord asked the tenant to agree to rectify the situation when the omission was discovered. This request was refused.

At the first hearing the judge made various findings of fact of which the most pertinent was that he considered that although the tenant's solicitors were aware of the omission that knowledge could not be implied to the tenant. RECTIFICATION was refused.

The landlord appealed to the Court of Appeal. The leading judgment stated: 'Here was a lease proffered by the landlord; it was repeatedly reviewed and reconsidered in the context of the very clauses ... which would be drawing the attention of the parties to rent and rent review; yet no suggestion was made of adjusting (the relevant) clause.' The Court of Appeal agreed with the decision of the court below that there was insufficient EVIDENCE as to unconscionable conduct on the part of the tenant's solicitor so as to impute knowledge to the tenant. The appeal was refused.

NO MACHINERY TO BE IMPLIED

General Accident Fire & Life Assurance Corporation plc v Electronic Data Processing Co plc [1987] 1 EGLR 112

The rent review clause identified eight assumptions that were to be taken into account by the arbitrator at the time of the rent review. The rent review machinery itself was not included in the list. The fact that the hypothetical term was for twenty years would mean that the absence of a rent review clause would result in a valuation which the arbitrator deemed to be fifteen per cent higher than if the provision were included. The tenant appealed. It argued that the commercial common-sense approach favoured by the Vice-Chancellor in *British Gas* [1986] 1 EGLR 120, should not permit the omission of that assumption to remain.

In *British Gas* the judgment included a request that litigation in the area of rent reviews should be restricted. This was a view disapproved of by Harman J, as he then was, in the instant (existing) case.

In any event Harman J held the parties to the bargain that had been agreed as to the terms. There would not be any machinery implied as an assumption.

MACHINERY FOR APPOINTMENT OF INDEPENDENT SURVEYOR

Darlington Borough Council v Waring & Gillow (Holdings) Ltd. [1988] 2 EGLR 159

There was a lease for a period of twenty years from 19 June 1978. Rent reviews were to take place every five years. There was a clause setting out the machinery by which an independent surveyor should be appointed.

> ... the council (the landlord) may during such period of six months before the commencement of the relevant period of five years (as to which time

shall be of the essence) require an independent surveyor ... to be appointed to determine the new rent.

The review was due in June 1983. Although correspondence passed between the parties the council did not request the President of the RICS to appoint until October 1984. The tenants did not participate. The landlords contended that the fact that the 'without prejudice' correspondence continued after the due date, with the tenants pursuing negotiations and indeed volunteering to agree a list or seek nomination of a surveyor, meant that the tenants were, by their acts, agreeing to an extension of the time limits strictly set out within the lease.

The judge found that there was no agreement to extend time for the appointment. Time had passed. The landlord, by seeking a nomination almost fifteen months after the due date, was too late. The rent would remain unchanged for the review period.

This was another decision that held the parties to their bargains.

MACHINERY FOR DETERMINATION OF RENTAL VALUE

Bissett v Marwin Securities Ltd. (1986) 281 EG 75
Here the lease set out various clauses for reviewing the rent. The first two clauses referred to various indices for which commencement dates were given; in the third clause the commencement date was omitted. This also gave the greatest rental increase. In addition, it mentioned 'a preceding rent review' whilst omitting any reference to 'or the commencement of the term'. In the first two clauses the commencement date clearly coincided with the commencement date of the term.

The tenant sought a declaration that, because the landlord had put forward the third (and most beneficial) option, if he failed to succeed on that clause by reason of the reference within that clause to a preceding rent review (which could not be possible at the first review) then he should not be allowed to fall back upon the first two clauses.

The court agreed that the third option was inoperable and refused to imply the phrase 'or the commencement of the term' into the clause, but allowed the landlord to fall back upon the first two clauses.

It should be noted that when the landlord served his trigger notice he did not refer to the alternative clauses upon which he was eventually forced to rely. He was fortunate in being allowed a second chance.

Trigger notices, where the review clauses contain alternatives, should set out all the alternatives.

RECTIFICATION OF THE RENT REVIEW PROVISION

The City of London Real Property Company Ltd. v CGU International Insurance plc (21 December 2000, Chancery Division, unreported)
An under-lease provided that '... the lessor shall give notice in writing to the lessees of its desire to vary the yearly rent payable hereunder as from the relevant review date....'

The sub-lessee requested the court to rectify the lease to reflect the agreement of the parties that the review should be upwards or downwards above the level of a threshold initial rent.

The court found that there was a mutual mistake and ordered that the sub-lease be rectified. There was substantial written and some oral EVIDENCE as to the thinking of the parties some years earlier, which more than supported the contention of the sub-lessee.

The machinery is now apparently effective for the implementation of the rent review. On occasions that machinery, or certain steps within that machinery, will be subject to a strict adherence to a timetable. The wording of such a time-table appears to cause the machinery to be ineffective if operated outside the time limits set down in the clause. However, the law seems to reject support for such rational behaviour and ignores the fact that both parties agreed to the implementation of the time limits which one party is now seeking to enforce. Here are some general rules.

Time of the Essence

This qualification was often inserted in various elements of the procedure in a rent review clause in an attempt to ensure that both the landlord and the tenant adhered to the timetable which was agreed between the parties at the outset of the lease. However, since the *United Scientific* case (see below) judicial decisions have generally been against any strict adherence to the rule. Surely it is time that a new clause were drafted whereby an effective procedure to benefit all parties could be set out?

It is felt in some quarters that, by the service of a properly drafted notice, one party can make time of the essence where originally time of the essence for a relevant clause did not exist. It is a procedure which should be treated with caution and only upon careful examination of the lease and possibly after obtaining counsel's opinion.

TIME OF THE ESSENCE – THE HOUSE OF LORDS' DECISION

United Scientific Holdings Ltd. v Burnley Borough Council [1977] EGD 195
The landlord appealed to the House of Lords to decide whether a time of the essence timetable was inflexible or not.

The House of Lords decided that if the clause *expressly* stated that each step of the procedure was required to be governed by a strict time to be of the essence limit, then that limit would apply. However, in the absence of such clear expressions, the courts should favour a lenient approach to a timetable. In this case the wording was insufficient to require strict adherence to a timetable.

Whether the House of Lords would make the same decision today is not certain. Two strong Courts of Appeal found that strict time limits should be adhered to, and that leniency should not be allowed!

TIME OF THE ESSENCE – CLEAR AND UNAMBIGUOUS CLAUSE

Monopro Ltd. v The Trustees of Boston University (7 August 1991, Queen's Bench Division, unreported)
The plaintiffs were landlords of five properties. These were let to the defendants upon substantially similar terms. The terms in question referred to the time which should be given by one party to the other as to termination of the agreement. Some agreements referred to the period as being during the month of September. The remaining agreements referred to the month of May. It was common ground (agreed) that the notices were not given during the particular months specified.

The landlord requested the court to declare that the phrases as to time were express and clear CONTRA-INDICATIONS, thus mandatory. The court examined *United Scientific* (see case above). The court decided that where there were express words such as in the instant case then those words were mandatory. The court granted the declaration.

This judgment is evidence against those who argue that drafting a foolproof and mandatory clause is next to impossible. This was effective drafting effectively enforced.

APPOINTMENT OF AN ARBITRATOR

The Trustees of Henry Smith's Charity v AWADA Trading and Promotion Services Limited [1984] 1 EGLR 116
In this case the tenant appealed against a finding that the express terms within which the rent review provisions were to be implemented were phrased in such a way as to be capable of a flexible interpretation. The terms in question set specific periods within which certain steps had to be taken. There was in addition a term which indicated the result of a failure to adhere to the time limits.

The landlord had attempted to seek the appointment of an arbitrator out of time. The court below had agreed and allowed the application.

The Court of Appeal found that the terms were sufficiently explicit as to be mandatory. The landlord's application was too late. The appeal of the tenant was allowed. This judgment was cited with approval in *Starmark* (see p. 26).

Pembroke St. Georges Ltd. v Cromwell Developments Ltd. [1991] 2 EGLR 129
The lease was to run for a period of ninety-nine years from June 1965. The first review of the rent took place after twenty-four years. There was a clause which defined the method for determining the rental value at the period of review. An arbitrator was to be appointed in the event of the parties failing to agree as to a rental level. The onus for the appointment was placed upon the landlords. The landlords were required to take positive steps to seek an independent DETERMINATION in the six months prior to 24 June 1989. This they failed to do.

No action was taken by the landlords until the latter part of November 1989. The tenants objected, saying that time was of the essence for the application for an arbitrator to be made; and as a result the landlord was too late. The counsel for the tenants submitted that, in spite of the indications laid down in *United Scientific* (see p. 14), time limits should be waived unless there were express and clear words to the contrary. Here there were clear words to the contrary. The clear statement in the lease as to when action should be taken was indicative of the parties' wishes to be bound by time limits.

The court disagreed. The time limits were no more than indicative. Therefore, the landlords obtained their review.

APPOINTMENT OF AN ARBITRATOR – THE SCOTTISH VIEW

Visionhire v Britel Fund Trustees Ltd. [1992] 1 EGLR 128 95
Here the Scottish Court of Appeal (i.e. the Inner House) was asked to decide two matters. First, did the *United Scientific* (see p. 14) guidelines apply when interpreting Scottish leases? If they did (more particularly the point concerning time limits), did the judgment affect the particular facts? The court decided that the English decision was as applicable to Scotland as to England and Wales.

In this case the landlords were required to make an application for the appointment of an arbitrator before the review date. If they failed to do so then the tenants could make a counter-proposal as to the valuation. If the landlords did not seek the appointment of an arbitrator within three months of the counter-proposal, the rent proposed by the tenants was the rent payable for the subsequent review period.

The landlords failed to meet both deadlines. They relied upon the *United Scientific* guidelines in the hope that they could apply out of time for the nomination of an arbitrator. The court rejected their appeal. To do otherwise would have changed the actual bargain to which the parties had agreed.

This approach seems eminently sensible and sets out what the parties actually decided to implement. It should be noted, however, that this decision contrasts with the *Pembroke* decision (see above) and others. It would be interesting to see what view the House of Lords determined were the matter of time of the essence put before them again.

TIME OF THE ESSENCE OF APPOINTMENT AND OF PUBLICATION OF AWARD

Kings (Estate Agents) Ltd. v Anderson and Another [1992] 1 EGLR 121
A letter of March 1990 from the landlord stated the level of the new rent. There was a valid counter-notice. The lease required that in the event of a disagreement an arbitration should take place.

The lease anticipated that the period of time for the rent to be agreed expired on 25 December 1989. Therefore, again relying upon the rent review clause, the tenant submitted that the time for appointment of an arbitrator had expired three months earlier.

In fact, the landlord required the RICS to appoint an arbitrator on or before 19 June 1990. However, the draughtsman of the clause had omitted a vital phrase in the review clause. The lease had reached the last or fourth phase. There was no mention of the fourth phase in the time of the essence qualification for the arbitrator to make his award. This time of the essence requirement had required a DETERMINATION to be made three months before the beginning of the review period.

Therefore, the court decided that there was no difficulty in concluding that time was not of the essence of this last review period. The landlord could therefore have his review in spite of being out of time. A possible action for negligence against the solicitor for the tenants?

APPOINTMENT OF A VALUER AND ISSUE OF A CERTIFICATE OF VALUE

Shuwa Ashdown House Corporation v Grayrigg Properties Ltd. and Another [1992] 46 EG 108

The lease review clause required the landlord to obtain a certificate of value within a certain period of time. There was a period of three months for attempts to agree the rental level. Should these attempts appear likely to be fruitless, then within that same period a certificate of value was to be obtained. Time was of the essence in this respect (whatever that meant).

Attempts to agree the rental level were made. The landlord failed to obtain the certificate within the specified period. The tenant considered that the landlord was prevented from implementing the review. The landlord disagreed and claimed that time being of the essence related to failure to agree and nothing more.

The learned judge found it difficult to support the tenant's view at a first impression. He relied upon a common-sense approach. This approach guided him to the opinion that it would be impossible for an independent valuer to certify within a three-month period, the reason being that the RICS was required to appoint. The appointment might be delayed, and the valuer would need to research the comparables. Subsequently, a certificate needed to be published. Therefore the three-month period was too short.

Having come to that conclusion, the judge decided that no competently advised landlord would have agreed to such a clause. In spite of deciding the matter in the favour of the landlord the judge concluded that the time of the essence phrase in fact added nothing to the review clause.

This conclusion does seem somewhat extraordinary. Here were two parties aware of the problems of timely action when they inserted a time of the essence clause into their contract. The phrase could only refer to the requirement for a certificate. The landlord would propose the draft terms of the lease to the tenant. The tenant would try to limit the period of delay likely before the certificate would be known, then it would be for the advisers of the landlord to alert him at the earliest possible juncture or be liable for professional negligence. With respect to the learned judge, it is suggested that commercial reality is frequently the basis of the agreement which the parties reach at arm's length and not a judge's view of the practicality of the procedure.

MACHINERY OF THE RENT REVIEW AND TIME OF THE ESSENCE

Power Securities (Manchester) Ltd. v Prudential Assurance Co Ltd. [1987] 1 EGLR 121

The premises were a shopping centre and, therefore, were separated into various units but leased as a whole by the landlord to the tenants. There were alternative elements in the machinery to assist the DETERMINATION of the reviewed rent. It was provided 'that the total income receivable for the demised premises shall be agreed between the landlord and lessee within six months after the expiration of the second year of the said term, and if not so agreed shall be the sum of £225,000.'

No move was made by either party to come to an agreement as to the total amount receivable, either by the date stated or for some time subsequently. The conclusion of the landlord was that the reviewed rent was to be £225,000 per annum. The tenant objected on the ground that it was not a DEEMING clause because time was not of the essence.

The court agreed with the tenant. The judge in this case attempted to determine principles relevant to the phrase 'time of the essence'.

1. 'The correct approach to a rent review clause is to begin with a presumption that time is not of the essence of the time limits laid down for the various steps to be taken for the determination of a revised rent or, it may be added, of any component element in its calculation.'
2. 'This presumption will be displaced if, on a consideration of the lease as a whole, and in particular of the provisions of the rent review clause as a whole, it appears that the parties have evinced a contrary intention.'
3. 'Where the parties have not only required a step to be taken within a specified time but have expressly provided for the consequences in case of default, this provides an indication of greater or less strength that time is to be of the essence, but it is not necessarily decisive. Whether it is so or not must depend on all the circumstances of the case, including the context and working of the provision, the degree of emphasis, the purpose and effect of the default clause and any other relevant consideration.'
4. 'In the end, the matter is one of impression to be derived from a consideration of the rent review clause as a whole, together with any other relevant considerations, avoiding fine distinctions, but giving effect to every provision in the lease.'

The court therefore decided that, since the parties might have anticipated that, in the event of failure to agree the value of the voids, a determination by an expert or arbitrator would be required, so it would be very unlikely that the parties could adhere to the six months' time limit as required. This being the case, there was a presumption against time being of the essence, and the literal construction put forward by the landlord was not accepted.

Trigger Notice

This is intended to be a form of words which gives a clear and concise instruction. The rent which the landlord seeks and the clause under which the notice is implemented should both be stated without ambiguity.

The phrases 'without prejudice' and 'subject to contract' should not head this notice or the counter-notice. However, it is essential to note that in *South Shropshire* (see p. 34) the Court of Appeal stated that privilege could apply even to an initial approach. This would not be wanted if there is a DEEMING clause.

Most cases concerning trigger notices stem from the question of their validity. It is generally considered that an oral notice is insufficient and ineffective; it should never be relied upon.

FORM OF TRIGGER NOTICE – APPLICABILITY OF COUNTER-NOTICE

Ian Yates and Others (1986) Outer House Cases Scotland
The landlord personally wrote an informal letter to the managing director of the tenant. The letter stated the level of the existing rent and set out the proposed increased rental level. This informal letter was informally acknowledged. Later the solicitors for the tenant sent a formal objection (counter-notice). The landlord sought a declaration that in spite of the informal tone of the initial letter that the clear reference to the revised rental level was sufficient to comply with the requirements of the lease.

The tenant was alleged to have failed to have submitted a formal counter-notice within the specified time limits. The landlord sought a further declaration that the proposed rent was the effective rent.

The judge, relying somewhat upon *Shirlcar* (see p. 37) and more upon *McCutcheon* [1964] 1 All ER 430, decided that where there was a possibility of doubt then the original notice was not effective. Here there was a possibility of ambiguity. The letter could have been a mere invitation to negotiate.

The court decided that as a result the tenants had not failed to serve an effective counter-notice formally opposing the rental proposal.

It is interesting to note that the court drew support from the decision of the Court of Appeal in *AWADA* (see p. 15) as well as the dissenting opinion in *Mecca* [1984] 49 P&CR 12, both of which have been endorsed in *Starmark* (see p. 26).

TIME OF THE ESSENCE – TRIGGER NOTICE – VALIDITY OF CONTENT

Commission for the New Towns v R. Levy & Co Ltd. [1990] 28 EG 119
The lease required the trigger notice to specify the proposed rent, a reasonable request. The notice did not do so. Time was of the essence for service of the notice.

The court decided that the phrase 'shall specify the yearly rent' was much more forceful than 'stating the suggested new rent to be reserved'. The requirement was mandatory.

The court was wise to rely upon the requirements of time being of the essence. To pin one's hopes as a professional adviser upon whether words are sufficiently forceful is too dangerous.

VALIDITY OF THE TRIGGER NOTICE

Durham City Estates Ltd. v Felicetti & Another [1990] 1 EGLR 143
A lack of proof reading (of which all are guilty) led to this appeal to the Court of Appeal. The trigger notice stated one sum in figures and another in writing.

The court asked itself whether the notice would have misled a tenant or its professional advisors.

The court found that the notice was an effective and clear notice. The court referred to a judgment of similar circumstances and there the sum written in letters was the sum chosen. In spite of the comment by Lloyd LJ that both the lease and the two pertinent letters were very carelessly drawn up, the court were unanimous in their decision.

The landlords were rather fortunate in that they were not penalised for the actions of their advisers. This decision is not a good precedent as far as encouraging a more professional attitude in this vital field.

TRIGGER NOTICE AND TIME NOT OF THE ESSENCE

Amherst v James Walker Goldsmith & Silversmith (1980) 254 EG 123
The landlord's agents served a trigger notice after the relevant date. The tenant argued that time was of the essence of the entire notice and valuation process. The court found that the landlord was not governed by any time limit for the service of the trigger notice. The tenant appealed. The Court of Appeal decided that the lease had explicitly made time to be of the essence for two out of three steps. However the procedure as to the service of the trigger notice was silent. The court refused to imply that time was of the essence where the parties deliberately omitted such a reference. The appeal was dismissed. This is the beginning of the series of decisions that properly hold parties to their bargains.

TIME OF THE ESSENCE AND TRIGGER NOTICES – CONTRADICTORY DIRECTIONS

Panavia Air Cargo Ltd. v Southend-on-Sea Borough Council [1988] 22 EG 82
There was a lease for a term of fifty-seven years. That time was not to be of the essence was stated as a generality and set out in paragraph 13. But in paragraph 14 was the following clause: 'If for any reason whatsoever no review of rent takes place in respect of any relevant period or a review is not completed within 12 months of the commencement of any relevant period then the rent

payable during the relevant period under the provision of this lease shall be increased by 25 per cent of the rent then currently payable.'

The point for decision was whether time was of the essence for the twelve-month period in spite of the clear wording in paragraph 13. The trigger notice was served before the commencement of the twelve-month period. Negotiations commenced. However, agreement was not achieved and the twelve-month period passed.

The Court of Appeal relied upon the *United Scientific* decision (see p. 14) to confirm that time should not be of the essence unless the contrary desire was expressed in very clear circumstances. Here the circumstances were not sufficiently clear to permit time to be of the essence and, therefore, the landlords were able to continue the review process.

TIME OF THE ESSENCE – TRIGGER NOTICE – *UNITED SCIENTIFIC* CLAUSE

McDonald's Property Co Ltd. v HSBC Bank plc [2001] 36 EG 181
The rent review provision was identical to that in *United Scientific* (see p. 14). The difference was that the landlords' notice was served out of time, by thirteen months. The tenant argued that time was of the essence for the service of the notice.

The court rejected the CONTRA-INDICATIONS as being insufficient. Time was not of the essence. Here was an excellent opportunity to distinguish the *obiter* comments in *United* and move to the stance exhibited in *Starmark* (see p. 26) of keeping the parties to their bargains.

TRIGGER NOTICE AND TIME OF THE ESSENCE

First Property Growth Partnership Ltd. v Royal Sun Alliance Property Services Ltd. [2003] 1 All ER 533
The trigger notice was required to be served 'at any time not more than twelve months before the expiration of ... every fifth year but not at any other time'. The lessor served the notice some thirteen months after the expiry of the allotted period. The landlord argued that time was not of the essence. There does not appear to be any explicit requirement that time be of the essence in any of the provisions.

The court decided that the notice must not be given more than twelve months before the expiration of the relevant period. The notice did not comply with the requirement that it should be given before the end of the five-year current period prior to the review. The court was supported in that view by the clear words 'but not at any other time' in conjunction with the service of the trigger notice.

Time was of the essence. The court must have decided these words were sufficiently clear CONTRA-INDICATIONS to avoid the need for express words.

This keeps parties to their bargain and must be lauded. However these conflicting judgments must cause despair to those attempting to advise whether to challenge or not.

TRIGGER NOTICES – TIME OF THE ESSENCE – SECOND CHANCE

London General Holdings Ltd. v Kingston Upon Thames Nominees Ltd.
[1993] 1 EGLR 133

The tenant asked for a declaration that the rent had not been reviewed at a certain date. The lease was for a term of thirty-five years. The landlord served a trigger notice. The notice did not mention any proposed rent. The notice had to be served not more than twelve months but not less than three months before the relevant date. The notice was in time. By reason of the contents of the trigger notice the tenant was not in a position to serve a counter-notice. However the tenant wrote explicitly asking for the proposed rental figure. There was no reply. A month later the tenant made an offer as to the revised rent. This offer was at once rejected. There was still no proposal. Two months later the landlord assigned its interest. At almost the same time the assignee served a trigger notice.

The lease had allowed for this situation. There was a fall back clause. If the landlord had failed to serve the trigger notice within the clearly specified period then the lease permitted one more opportunity to serve the notice at any time.

The court found that the rent could still be reviewed.

Suitable adjectives are not easily found to express sufficient sympathy with the tenant. However a very shrewd and wise fall back clause saved the assignee. Tenants in a strong bargaining position might be tempted to direct that the addition be removed.

TWO TRIGGER NOTICES – CONFUSION

Cordon Bleu Freezer Food Centres Ltd. v Marbleace Ltd. [1987] 2 EGLR 143
The freeholder (lessor) sold his interest. There was a rent review imminent. Time was of the essence. The managing agents, after the date of the sale, sent a trigger notice to the tenants quoting in the notice the name of the original freeholder and a rent of £45,000. The tenants queried the validity of the notice. Subsequently, the managing agents were appointed to manage the property by the new owner. However, the solicitors for the new owner had sent to the tenants what purported to be a trigger notice for the same review but quoting a rent of £30,000 per annum. Hence the confusion. The tenant, not surprisingly, concluded that the trigger notice from the managing agent quoting the incorrect lessor was invalid. The new landlord disagreed.

The tenant served a counter-notice in time to the effect that the rent should be £16,000. The counter-notice was a DEEMING notice and, as served in time, was also effective. The court decided that the landlord had failed.

TRIGGER NOTICE – VALID SERVICE OR NOT – MEANING OF 'THE PERIOD IN QUESTION'

Maraday Ltd. v Sturt Properties Ltd. [1988] 46 EG 99
The demised premises were subject to a lease of fifteen years. The first review commenced on 25 March 1987. The notice was required to be served at least

six months before the expiry of the period in question. The trigger notice was served on 26 January 1987.

The tenants claimed the notice was out of time as it should have been served before 24 September 1986. The landlords disagreed. They claimed the word 'expiry' meant 'end' and therefore they could serve their trigger notice at any time up until 23 October 1991.

The learned judge could not see how he could give the meaning 'commencement' to the word 'expiry' and therefore agreed with the landlords that the notice was good.

TRIGGER NOTICE – TIME OF THE ESSENCE – VALID SERVICE OR NOT – INEFFECTIVE DELIVERY

Stephenson & Son v Orca Properties Ltd. [1989] 2 EGLR 129
The period for the review of the rent expired on 30 June 1985. The landlords sent a trigger notice by recorded delivery post, which was capable of being delivered on 29 June 1985. However on that day, a Saturday, there were no staff at the office. Therefore the postman was not able to effect delivery and obtain a receipt until Monday thereafter. Monday was outside the period for review.

The court decided that the notice was not delivered in time and was not deemed to have been delivered on the Saturday. Therefore the trigger notice was out of time.

This was exactly the correct but harsh decision. Time was of the essence. There can be no sympathy for the landlord for allowing the service to take place at the last minute.

TRIGGER NOTICE – WHERE TIME WOULD BE OF THE ESSENCE

Wing Crawford Holdings Ltd. v Lion Corporation Ltd. [1989] 1 NZLR 562
The trigger notice was served almost nineteen months after the due date for rental revaluation. The tenant argued that the notice was out of time as time was of the essence. The court decided that the rent review provisions did not contain any explicit indication that time was to be of the essence. The court could not draw any inferences from the surrounding circumstances. Time was not of the essence. The service of the notice was valid.

The court stated that the tenant could have taken the steps required to trigger the review and also make time of the essence.

The judge had regard to the three categories of cases identified in Woodfall's *Law of Landlord and Tenant*[*] where time would be of the essence: where the landlord's right to require arbitration as to the new value was to be exercised within a certain period and not otherwise; when the rent review was explicitly interlocked with the lease BREAK CLAUSE; and finally where time was explicitly stated to be of the essence or was intended to be so. The judge identified the third category as applicable. However the facts prevented time being of the essence. The rent was revalued.

* (28th edn., Sweet & Maxwell)

TRIGGER NOTICES AND DELAY

Telegraph Properties (Securities) Ltd. v Courtaulds Ltd. (1980) 257 EG 1153
A period of five and three-quarter years passed before the landlord served a trigger notice. The High Court decided that even where time was not of the essence that delay was too long. The service was ineffective.

This decision has been disapproved in an *obiter* judgment of Oliver LJ in *Amherst v James Walker Goldsmith & Silversmith* (1980) 254 EG 123. There the Court of Appeal held that a delay of four years where time was not of the essence did not prevent the trigger notice being served.

Once more there is a conflict between the commercial and the literal approach. The former indicates that far too long has passed for it to be either practical or fair to initiate proceedings. This must be preferable to the alternative approach, which relies upon the strict literal interpretation of the clause and, therefore, permits a landlord to issue a trigger notice whenever he wishes, subject only to a restrictive phrase of total clarity that time is to be of the essence.

TRIGGER NOTICE – YEARS LATE AND WAS VALID. DATE OF VALUATION, DATE OF REVIEW OR DETERMINATION

Glofield Properties Ltd. v Morley and Another (no. 2) [1989] 2 EGLR 118
An independent surveyor had been appointed to determine the revised rental value. The DETERMINATION had not been made. The landlord suggested that the comparable EVIDENCE should indicate the market value at the date when the determination was made. The tenants argued that the appropriate date was the date when the review was made.

The Court of Appeal allowed the appeal of the tenants. The court decided that the concept of a rent review was to determine the value as at the date of the commencement of the period in question, not at some period significantly later in time.

In spite of the fact that the date of valuation was described as being at the date when the determination was made, the court gave an equitable interpretation to 'determination'.

Again it could be suggested that the court overrode the bargain into which the parties freely entered. A fair but possibly unjust decision.

TRIGGER NOTICE – CAN LANDLORD BE FORCED TO SERVE? YES

Addin v Secretary of State for the Environment [1997] 1 EGLR 99
Here only the landlord was permitted to serve the trigger notice. At the time of the review the rental values had fallen. The tenant wished to take advantage of the review provision, which was not upward only. The landlord had no interest in taking action, as there was a passing rent well in excess of the current value. The tenant asked the court to order the service of a trigger notice.

The court noted that the provision used only the word 'until' the revised rent had been agreed. The words were not 'unless and until'. This, the court implied, meant that a revision was inevitable.

The court assumed that at the time that the lease was prepared the parties were deemed to be men of commercial substance and alive to the ways of landlords and tenants.

The judge found that the clause meant that at each review period the tenant shall pay the higher of either the base rent or the current value. The landlord was in effect ordered to have a review.

This is a very strange situation. As the court said, the parties were alive to the realities. Thus they would have been aware of the possibility of the current situation. This decision will have disastrous consequences for funding of properties where there is a landlord-only trigger notice. There is no debate with the reasoning of the court as to the word 'until'. However to rewrite the bargain to assist the tenant is very unfortunate.

TRIGGER NOTICE – CAN LANDLORD BE FORCED TO SERVE? NO

Sunflower Services Ltd. v Unisys New Zealand Ltd. [1997] 1 NZLR 385 Privy Council
The lease permitted only the landlord to trigger the rent review process. The property values had fallen. The landlord refused to comply as was its right.

The Privy Council noted that the provision used the word 'may' and not 'shall'. Apparently there were no indicative phrases such as 'until' and if there were they did not impress the Privy Council.

A unanimous recommendation delivered by the eminent contributor to rent review law, Lord Browne-Wilkinson, allowed the appeal of the landlord against a direction by the Court of Appeal of New Zealand that the landlord implement the review process.

The comparison between this judgment and *Addin* (see case above) is fortuitous but instructive. The sooner a landlord takes a similar situation to the House of Lords the better so that the situation can be finally resolved.

Counter-Notice

This is a notice sent by the tenant in reply to a trigger notice sent by the landlord. It need state only two matters: that the tenant disagrees with the rent sought in the trigger notice and that the issue (the amount of rent that should apply under the review procedure) should be referred to an arbitrator or an independent expert, whichever applies in that instance.

ORAL COUNTER-NOTICE

Museprime Properties Ltd. v Adhill Properties Ltd. [1990] 36 EG 114
The dispute arose out of an auction of three shop units with residential upper parts. Trigger notices were served correctly and in time. Two of the tenants

objected orally. There was no written confirmation of these objections from either party. The auction details and later announcements referred to counter proposals by the tenants. As to two of these counter proposals there was no EVIDENCE at all. The purchasers applied to rescind the contracts on the ground of misrepresentation.

The court rescinded the contracts.

Only written counter-notices could be effective. This is obvious but needed stating here. There was no reason for the landlord, or his agent, not to confirm the contents of telephone conversations with the tenants. Whether these notional confirmations would have been sufficient to be deemed effective counter-notices is debatable. However silence is not golden where money is concerned.

Failure to Serve Counter-Notice in Time

In *Bellinger v South London Stationers Ltd.* (1979) 252 EG 699, and subsequently in *Oldschool v Johns* (1980) 256 EG 381, there was a rent review clause the procedure of which required adherence to strict time limits. The landlord observed the correct time limit for service of the initial trigger notice. The tenant's letter of acknowledgement and dispute of the initial rent was not deemed to be of such effect as to be a proper counter-notice. In that event the court decided that by reason of the wording of the clause the rental figure quoted in the initial trigger notice should be the rent of the review period.

COUNTER-NOTICE AND TIME OF THE ESSENCE

Factory Holdings Group Ltd. v Leboff International Ltd. (1986) 282 EG 1005
The landlord's trigger notice was served in accordance with the terms of the lease. The tenant served a notice, making time of the essence for the REFERENCE to arbitration, such a reference to be made by the landlord within twenty-eight days. The tenant sought to rely on this notice when the landlord failed to comply.

The judgment was for the landlord. The court indicated that once it is accepted that there is a presumption against time being of the essence, then it is required to look at each particular instance to ascertain whether the presumption has been negated by express words.

1. A tenant may serve a notice upon the landlord making time of the essence, when the landlord is not taking any action.
2. However, the tenant is not entitled to serve a notice making time of the essence where the action he requires the landlord to take could be taken by the tenant himself, i.e., the reference to arbitration.

TIME OF THE ESSENCE AND SERVICE OF THE COUNTER-NOTICE

Starmark Enterprises Ltd. v CPL Enterprises Ltd. [2002] CH 306
There was a timetable which set out the period within which the tenant should serve a valid counter-notice. The clause stated that if the counter-notice were

not served within the stated period then the rental figure proposed in the trigger notice would become the actual rent. The counter-notice was served outside the actual period specified.

The tenant successfully opposed the imposition of the proposed rent by relying upon the argument that time was not of the essence unless expressly stated to be so. The landlord appealed.

The Court of Appeal considered that the words used were sufficient to be CONTRA-INDICATIONS that time should be deemed to be of the essence. It was decided that the parties should be bound by their bargain.

An earlier decision of the Court of Appeal, more favourable to the tenant, was distinguished. Further the Court of Appeal drew support from the strict approach exhibited in Scotland, *Visionhire* [1992] 1 EGLR 128; Australia, *Mailman (GR) & Associates Pty* [1991] 24 NSWLR 80, and New Zealand, *Mobile Oil New Zealand Ltd.* [1995] 3 NZLR 114.

The appeal was allowed. The proposed rental figure became the actual rent for the review period.

VALIDITY OF TRIGGER NOTICES AND COUNTER-NOTICES AND WHETHER TIME OF THE ESSENCE EFFECTIVE

Norwich Union Life Insurance Society v Sketchley plc (1986) 280 EG 773
Part of the rent review machinery stated that, 'The landlord shall be entitled by notice in writing given to the tenant at any one time...'. The timetable in this matter was as follows:

An initial letter was sent on 4 August 1982 by the landlord's agent but to the wrong address.
A reply was dispatched by the tenants querying whether the notice was addressed to them.
A subsequent letter from the agents, dated 27 August 1982, was sent, correctly addressed.

The tenant claimed that any attempt to initiate this procedure was out of time. Not until June 1983 did the solicitors for the landlord attempt to rectify the matter by sending what was termed a formal notice to the tenant. But that letter acknowledged that the letter of 27 August 1982 was intended to be a formal notice.

On 8 November 1983, the landlord served a notice dependent on the letter of 27 August 1982 seeking the appointment of an expert to determine the value of the rent in default of agreement. Whether that letter was itself out of time was another question for the court.

Judgment was given for the tenant. Notice, as served, was sufficient for the purposes of the lease. The court stated that a notice is intended to give information. If it does succeed in imparting the requisite information then it can properly be described as a notice in writing. This notice was a good notice. It was also accepted as such by the tenants in their reply querying whether the letter was actually for them.

As to whether the landlord was able to serve a second trigger notice, the court decided that the rent review procedure could only be initiated once.

TIME OF THE ESSENCE – COUNTER-NOTICE

Legal and Commercial Properties Ltd. v Lothian Regional Council [1988] SCLR 201

The terms of the rent review provision gave the tenants twenty-eight days within which to respond to the trigger notice. If there were no response then the proposed rent became effective. The trigger notice was served on 8 October 1986. The tenants formally responded upon 14 November 1986. The counter-notice was not served within the twenty-eight day period.

The court noted the decision in *AWADA* (see p. 15) and the dissenting decision in *Mecca* [1984] 49 P&CR 12 and the decision in *Yates* (see p. 19). These decisions were to the effect that the phraseology of the directions can be sufficiently explicit to indicate that time was of the essence.

However, in this case, the court examined the detail of the provision and decided that time was not of the essence of that particular element.

The court was persuaded by the absence of time restrictions for the request for the appointment of an independent surveyor.

There are comments that where timetables place impossible goals upon parties then leniency should be extended. It is not for courts to change the bargains of the parties, the confidential detail of which will never emerge and cannot in any event influence interpretation.

It is for the parties to give explicit instructions and ensure that they are carried out.

TIME NOT OF THE ESSENCE FOR FRESH TRIGGER NOTICE

Saloma Pty Ltd. v Big Country Developments Pty Ltd. (15 December 1997, New South Wales Supreme Court, unreported)

The landlord served a trigger notice. The tenant noted that the notice indicated an increase in the monthly amount of rent and outgoings. The review provision required that the trigger notice state an increase in the annual market rent. Time was not of the essence for service of the trigger notice. Time was of the essence for the service of the counter-notice.

The notice did not state the increase in the format that was required by the provision.

The judge found that the trigger notice was very deficient in that the tenant could not possibly determine whether to accept the proposal or not from the way the information was provided.

The trigger notice was invalid. The tenant was not penalised. The landlord could serve a fresh trigger notice, as time was not of the essence of the first stage.

TIME OF THE ESSENCE – COUNTER-NOTICE – EFFECT OF CONTINUING
NEGOTIATIONS

W. & R. Jack Ltd. v Fifield [1999] 1 NZLR 48 and [2001] LTR 39 Privy
Council
The tenant sought an order appointing an arbitrator. The court refused. The
tenant appealed. The rent review clause was upward only. There were time
limits for both the service of the counter-notice and the appointment of the
arbitrator. The tenant served its counter-notice late by two and a half months.
In that notice the tenant rejected a compromise offer of the landlord. The tenant sought an extension of time for the appointment under the Arbitration Act
of New Zealand. It also raised ISSUES of WAIVER and ESTOPPEL against the
landlord. The Court of Appeal directed itself solely to the extension issue. It
found that the delay was only twelve weeks and there would be financial hardship if the deemed rent became the actual rent. The hardship was out of
proportion to the fault. The tenant was granted the extension,
     The landlord appealed to the Privy Council. The Privy Council recom-
mended against granting the appeal.
     It strongly rejected the extension of time argument. It found that the action
of the landlord by continuing negotiations and then attempting to implement
the DEEMING provision was sufficient to exempt the tenant from any liability.
     This refreshing practical decision does not in any way relieve all parties of
the absolute need to comply with the time limits and serve notices. Negotiations will still continue and if they do not the representatives who take offence
are not professional.
     If there is any doubt, written confirmation that the time limits will not be
adhered to should be obtained by recorded delivery or other formal method.

TRIGGER AND COUNTER-NOTICES – TIME OF THE ESSENCE VALIDITY

Fox & Widley v Guram & Another [1998] 1 EGLR 91
The trigger notice was served. The notice stated the date of the review
incorrectly by a few days. The court decided that no one was misled. The
notice was valid.
     The tenants then argued that it was not evident that the agents who served
the notice had authority from the landlord to take that action. The court
decided that although the affidavits were silent on the point, as the agents had
represented the landlords for some years it was reasonable to infer that they
had authority.
     The review provisions required that the landlord have the open market
rental value determined at the time of the trigger notice. The tenant argued that
there was no DETERMINATION or if there were it was not genuine. The court
ruled that there was no requirement for the value estimate to be genuine.
     This last decision causes concern. The requirement was for a determina-
tion. That is not an estimate. A determination should have been made. The
phraseology is unfortunate for the landlord, but that is the term used.

The tenants had replied to the trigger notice without making an objection, merely seeking an explanation. Time was of the essence and time had expired. The tenants made an application for an extension of time to commence arbitration proceedings. The court refused. The extension should be given in very exceptional circumstances and a windfall to the landlord was not one of them.

Extensions of time for the commencement of arbitration proceedings will not succeed very often, if at all, when based upon the old excuses.

COUNTER-NOTICE – TIME OF THE ESSENCE COMPLIANCE

Scottish Life Assurance Co v Agfa-Gevaert Ltd. [1998] SCLR 238

The tenant appealed against a decision that the form of its counter-notice had not complied with the review provisions. The counter-notice was required to be served within one month of receiving the trigger notice. Also the counter-notice was required to state the tenant's estimate of the rental value. It did not do so.

The landlord had sought a declaration that the new rent was the proposed rent as a result of a DEEMING provision. The court below granted that declaration.

The tenants argued that the requirements for the counter-notice were directory not mandatory. The court found that the terms were mandatory. Here, unlike *Patel and Another v Earlspring Properties Ltd.* (see p. 32), there was no body of correspondence which appeared to accept the counter-notice as valid. The court found that the provision did not lack common sense nor was it inconsequential.

The decision supports the basic attempt of the draftsman to impose a structure upon the procedure.

Whether the English courts will follow the more practical approach of the Scottish courts remains to be seen.

TIME OF THE ESSENCE WITH MORE THAN ONE TRIGGER NOTICE

Patel and Others v Peel Investments (South) Ltd. [1992] 30 EG 88

A shop property in Croydon was leased for twenty-eight years from 29 September 1983. There were five-yearly reviews. There was a complicated review procedure. Time was to be of the essence of one element. That element was that the tenant must apply for an arbitration to take place if the tenant disagreed with the rental value within three months of the service of the trigger notice.

Here confusion was rife. The landlord served a notice on 7 July 1988 stating that £19,500 was the new rental valuation. No counter-notice was served by the tenant. In the meantime, the landlord was taken over. Staff left and there was no correspondence on the files. The landlord thus served another trigger notice on 15 August 1989. This time the valuation was for £23,000. The agents for the tenants served what they considered to be a counter-notice on 8 September and made various ironic comments concerning duplication and incompetence.

The landlords, learning about the earlier notice and realising that there had not been any counter-notice to that notice, sought to rely upon that original valuation as binding. The tenants objected on the grounds of EQUITY. They claimed that the landlords had either waived their rights, or that they were estopped from relying upon the earlier notice, or that they had abandoned their original notice. The learned judge decided that the WAIVER point failed. The ESTOPPEL point and the abandon point failed as well. The landlord was successful at stage one. However, the tenant SUCCEEDED ON TERMS to have an extension of time granted for an application for the matter to be determined by an arbitrator.

FORM OF NOTICE AND COUNTER-NOTICE

Amalgamated Estates Ltd. v Joystretch Manufacturing Ltd. (1980) 257 EG 489

There was a lease in which the rent was to be reviewed every five years. The machinery for the review was clear and the service of the notice and counter-notice was subject to strict time to be of the essence qualifications. The trigger notice was served as was the counter-notice. However, the counter-notice omitted to seek the DETERMINATION of the rent by REFERENCE to arbitration; it merely stated the proposed rent to be too high. The landlord succeeded in obtaining the new rental figure by means of summary judgment. The tenant appealed.

The court agreed with the landlord.

1. The rent specified in a notice is not required by implication to be reasonably capable of being the open market rent.
2. The validity of such a notice is supported even where the timing of the notice is such that it would be impossible for the tenant to reply within the period set down in the clause even when time is of the essence.
3. The tenant is always able to apply for an extension of time to serve a counter-notice under s. 27 of the Arbitration Act 1950. This extension is unlikely to be refused when the facts in the clause above are recited.
4. The counter-notice must state not only that the quoted rent is disputed but that the dispute should be referred to arbitration.

Barrett Estate Services Ltd. v David Greig (Retail) Ltd. [1991] 36 EG 155

A trigger notice was served effectively. The counter-notice was required to state that the tenants did not agree with the value proposed by the landlord. That appeared to be sufficient.

The letter itself was not apparently headed 'counter-notice'. It merely said: 'I note that your assessment of rental value is £190,000 per annum, which I consider to be excessive. The review provisions are somewhat complicated and allow your client to assess the rental on the greater of two different bases.... I look forward to hearing from you or your clients' surveyors, in due course.'

There were subsequent discussions about negotiations. The landlord tried to insist that no counter-notice had been served. The court found the letter was quite sufficient.

It is somewhat strange as to why after extensive correspondence in the journals, professionals still fail to rely upon clear, explicit language in counter-notices. The cost of such incompetence in monetary terms is high and totally unnecessary.

Prudential Property Services Ltd. v Capital Land Holdings Ltd. [1993] 15 EG 147

The trigger notice was served in time. There were twenty-eight days for the service of a counter-notice. The response, sent in time, indicated that it was a formal notice and that the tenants disagreed as to the rent proposed.

The landlords did not treat this letter as a counter-notice. The court decided that there was a letter in writing and that it was sent in time.

Once more the problem was caused by apparent over-officiousness by the landlord or their representatives and by modest verbal laxity on behalf of the tenants. Another case of unnecessary cost.

Patel and Another v Earlspring Properties Ltd. [1991] 2 EGLR 131

A formal notice was served by the landlord in time. The letter from the tenant acknowledged the notice. However, it pleaded that the business could not justify the level of rent proposed and sought a reconsideration of the rental level.

The landlord acknowledged the letter and promised to respond soon. No response came until some months later when the tenants were told that they were in arrears of rent. The arrears arose as a result of the landlord charging the tenant the new rent which the tenant had not paid.

The tenant not surprisingly objected and asked the court to clarify the matter. The judge of the county court supported the tenant. The landlord appealed and failed to persuade the Court of Appeal. The counter-notice was effective.

There was an additional point. The tenant said that the landlord had failed to negotiate and therefore had lost its right to have a review in any event. The Court of Appeal disagreed. There was no need to go through a charade of renegotiation merely in order to comply with the wording of the clause.

As a result it is wise for the landlords to state in clear language that there is a requirement to negotiate, that they are willing to listen to offers from the tenant, but as to the valuation they are satisfied as to the suggested levels.

Privilege

Professional men and women frequently forget when the phrases 'without prejudice' and 'subject to contract' should be used. This section tries to clarify matters.

Public Privilege

Public privilege is not an area which will be encountered frequently in the rent review process.

Private Privilege

Private privilege is basically divided into two areas. The first area is correspondence between solicitor and client. Such correspondence is always privileged, but the privilege and protection may be waived if the person relying on that privilege no longer seeks to rely on it. Thus, it is quite in order for the party who wrote to the solicitor to put that correspondence into the hearing should it be so desired. It is the other side who cannot force the parties to the correspondence to make it public.

The more disputed area is where instructions are given to a third party to prepare a report (or valuation) when a dispute has arisen and arbitration or litigation is envisaged. It is the author's view that the proposed rent will either be accepted without demur, in which case no dispute will arise, or the proposed rent will be disputed. In the latter case, a dispute can automatically be deemed to exist, in which event the initial report of the landlord's surveyors to determine a proposed rent upon review should be protected. It is argued by some surveyors and lawyers that not all initial proposed rents upon review are disputed by tenants and that the privilege, which attaches when a dispute is in contemplation, should not therefore apply in the instance of an initial valuation.

Without Prejudice

The legal system of England and Wales is very keen to encourage amicable settlement of disputes between parties. Among the areas of correspondence which private privilege is intended to protect therefore are those between parties attempting to procure settlement of a dispute. Frequently these letters are headed 'without prejudice', with the purpose of trying to prevent the use of that letter in court or arbitration should the matter of dispute not be settled by the parties.

As previously mentioned, a heading such as 'without prejudice' should never apply to the initial notice from the landlord or to the counter-notice from the tenant, since both sides are making clear statements of their position. The landlord is stating categorically what he deems to be the relevant rental value which applies to the premises. The tenant also wishes the world to be clear as to his position that he does not agree with the rent quoted, for whatever reason. Subsequent correspondence between the landlord and tenant, prior to or in conjunction with a reference to an arbitrator or expert, may be headed 'without prejudice' if the parties are attempting to come to an agreement over the new rent. Should these attempts fail, the landlord will not want the arbitrator or expert to know how low a figure he would actually accept, nor, conversely, would the tenant wish to reveal how high he is prepared to go.

'WITHOUT PREJUDICE' CORRESPONDENCE

South Shropshire District Council v Amos (1986) 280 EG 635
In this case the Court of Appeal issued guidelines for obtaining effective
protection for 'without prejudice' correspondence.

Under planning law, a discontinuance order had been served. At the same
time, a general letter had been sent by the agents for the claimant indicating a
wish to negotiate and accepting the existence of a claim. The claim was
described as being such 'as will be full and final under all heads…'. The local
authority wished subsequently to introduce the letter into proceedings. The
claimant objected. The Court of Appeal confirmed that the letter was
privileged.

To obtain the protection of 'without prejudice' correspondence:

1. There must be an initial express intent to negotiate. This does not necessarily occur in trigger notices issued under a rent review clause, although
   convincing evidence to the contrary has yet to be produced. Thus the need
   to ensure such notices are not issued on a 'without prejudice' basis is still
   essential until there is a decision to the contrary.
2. There must be a dispute in existence. It appears that here the Court of
   Appeal takes a very wide view of what is a dispute.
3. The document must be marked 'without prejudice'.

To quote the judgment:

> in order to avoid any possibility of future unnecessary disputes about such
> matters we conclude by stating that we agree with the learned judge (a) that
> the heading 'without prejudice' does not conclusively or automatically
> render a document so marked privileged; (b) that, if privilege is claimed but
> challenged, the court can look at a document so headed in order to deter-
> mine its nature; and (c) that privilege can attach to a document headed
> 'without prejudice' even if it is an opening shot. This rule is, however, not
> limited to documents which are offers. It attaches to all documents which
> are marked 'without prejudice' and form part of negotiations, whether or
> not they are themselves offers, unless the privilege is defeated on some
> other grounds as was the case in *Re Daintry* [1893] 2 QB 116.

This ruling stated that an opening letter should not be headed 'without prejudice' if it were intended to be effective. The Court of Appeal disagreed with
this sensible action, which is particularly appropriate to a trigger notice.

The circumstances described in the present case do not equate in any way
with a trigger notice, which must be clear and unequivocal. At the time of the
letter, the District Council had not disputed the figures as it could not – it had
not received them. Therefore, apparently, the Court of Appeal was saying that
a dispute was *inevitable* in spite of the fact that the District Council could not
dispute the claim as it had not received it. Such opening correspondence was

therefore permitted to be without prejudice. Could this mean that a trigger notice could be served without specifying any rent and separately a 'without prejudice' letter sent specifying the requested rent?

Rush & Tompkins v Greater London Council [1989] AC 1280 House of Lords
The judgment of the House of Lords given by Lord Griffiths on behalf of their Lordships provides useful clarification:

> That the 'without prejudice rule' is a rule governing the ADMISSIBILITY of EVIDENCE and is founded upon the public policy of encouraging litigants to settle their differences rather than litigate them to a finish.

This concept of early settlement should be kept at the forefront of all involved in the rent review process.

> There is also authority for the proposition that the admission of an 'independent fact' in no way connected with the merits of the cause is admissible even if made in the course of negotiations for a settlement.

This is a very important comment as frequently matters are placed within a 'without prejudice' document that are totally outside the sphere of negotiations. The question is then how to produce that matter without risking sanction. Arbitrators and experts must be alert to this situation.

His Lordship could not accept the view of the Court of Appeal that *Walker v Wilsher* [1889] 23 QBD 335, was authority for the proposition that, if the negotiations succeed and a settlement is concluded, the privilege goes, having served its purpose.

His Lordship held that such a concept of disclosure would surely be equally discouraging if the main contractor, one of the protagonists in the case, knew that if he achieved a settlement those admissions could then be used against him by any other subcontractor with whom he might also be in dispute.

This comment applies equally to a landlord negotiating with a variety of tenants in a situation where concessions may be made in order to retain the attraction of the location.

'WITHOUT PREJUDICE' GUIDELINES

These guidelines are taken from the Court of Appeal judgment by Balcombe LJ in *Rush & Tompkins* [1988] 2 WLR 533, [1988] 1 All ER 549.

> ... the case has disclosed what appear to be some widespread misconceptions as to the nature of 'without prejudice' privilege. In an attempt to remove those misconceptions, and to give guidance to the profession, we venture to state the following principles.

(1) The purpose of 'without prejudice' privilege is to enable parties to negotiate without risk of their proposals being used against them if the negotiations fail....

(2) It is possible for the parties to use a form of words which will enable the 'without prejudice' correspondence to be referred to, even though no concluded settlement is reached....

(4) The privilege does not depend on the existence of proceedings.

(5) Even while the privilege subsists, i.e. before any settlement is reached, there are a number of real or apparent exceptions to the privilege. Thus: (a) the court may always look at a document marked 'without prejudice' and its contents for the purposes of deciding its ADMISSIBILITY.... This is not a real exception to the privilege, since the court must always be able to rule on the admissibility of a document, when a claim to privilege is challenged. It is under this head that the court can look at the documents to see, e.g. if an agreement has been concluded and, if so, to construe its terms; (b) the rule has no application to a document which, by its nature, may prejudice the person to whom it is addressed ... (c) there may be other exceptions (see Phipson [on Evidence (13th edn., 1982)], para. 19-11) but we do not think it appropriate to consider them further, since they do not arise in the context of the present case.'

There was comment critical of this statement by the House of Lords who are not in favour of widening the exceptions.

(6) The privilege extends to the solicitors of the parties to the 'without prejudice' negotiations.... However, we do not think it necessary or desirable to express any view on the question whether the privilege is valid against a third party (other than a party's solicitor) when no settlement has been reached by the parties to the 'without prejudice' negotiations.

This recommendation appears to receive support from Lord Griffiths.

'SUBJECT TO CONTRACT' AND 'WITHOUT PREJUDICE'

Henderson Group plc v Superabbey Ltd. [1988] 2 EGLR 155

The rent review was due. The agent for the landlord visited the premises. On a later date he sent a letter to the tenants. The letter was headed 'subject to contract' and 'without prejudice'.

The tenant responded unequivocally that the rental value was acceptable. A few days later the landlord wrote to the tenant enclosing a memorandum purporting to set out the terms of agreement, for signature by the tenant. This was not signed by the tenant, who wrote six weeks later saying that the new rental level was not acceptable. The landlord considered that the correspondence bound the tenants. The tenants disagreed.

The judge in a very clear reasoning set out various propositions as to the relevance of 'subject to contract':

1. Negotiations commenced under the heading 'subject to contract' retain that conditional form until the agreement is made.
2. It is the actual letter which is relevant, not the thought processes surrounding it.
3. The phrase 'subject to contract' means that the parties to the negotiations do not intend to be bound until an agreement has been signed which sets out all the terms and has been signed by the parties.
4. The conditional nature of the negotiations do not cease to be conditional once the parties have merely agreed upon the terms – such negotiations are only made unconditional by a positive act to that effect.
5. It is very unusual for the conditional meaning placed upon the phrase to be able to be removed.
6. In each circumstance, the actual intent of the parties evidenced from the document must be construed to determine whether the requirement of a conclusive contract is an essential element of the transaction or a mere whim of the parties.

The landlord argued that the use of the phrase 'subject to contract' meant nothing – being merely a mistake made frequently by commercial surveyors who have no understanding of its meaning. The judge disagreed. He considered that surveyors are well aware of its meaning. The use of the word 'proposed' supported the view that the landlord's agent did not intend to do more than merely alert the tenant that negotiations should, or could, start.

The judge decided that the exchange of letters was not binding upon the tenants.

'SUBJECT TO CONTRACT' – EFFECT ON TRIGGER NOTICE

Shirlcar Properties Ltd. v Heinitz (1983) 268 EG 362

Here a clear trigger notice of the rent required upon review was served, but the notice was headed by the phrase 'subject to contract'. When the landlord tried to rely upon this notice, the tenant disputed its validity.

The Court of Appeal gave judgment for the tenant.

1. The Court of Appeal decided that the notice must be certain and the phrase 'subject to contract' negated certainty.
2. The court considered that a reasonably minded tenant would consider that the phrase indicated that the notice was not binding but merely an invitation to negotiate.

If this decision were not correct, which logic indicates it must be, it would give an opportunity to the landlord to ascertain the market EVIDENCE likely to be produced by the tenant, and subsequently to serve a binding notice on the tenant at a rent which was at the top of the range of the evidence produced by the tenant. The tenant would be prevented from claiming the protection of privilege when DISCOVERY of that evidence was sought.

'SUBJECT TO CONTRACT' AND 'WITHOUT PREJUDICE' – TRIGGER NOTICE – TIME OF THE ESSENCE

Royal Life Insurance v Philips [1990] 43 EG 70

The trigger notice was another example of a last minute action, which sadly was not penalised. The notice was served four days before the expiry of the period for service, of which time was of the essence.

In spite of clear advice from various commentators the notice was headed 'subject to contract' and 'without prejudice'. Those indications should have nullified the notice.

The court decided that a reasonably minded tenant would have appreciated that the notice was an unequivocal trigger notice and responded appropriately. The notice was valid.

An extraordinary judgment, which professionals and landlords should treat with caution.

'SUBJECT TO CONTRACT'

Westway Homes Ltd. v Moores and Others [1991] 2 EGLR 193

Here a notice purporting to exercise an option to purchase was headed 'subject to contract'. The recipient argued that it was equivocal.

The Court of Appeal found that the heading was meaningless and of no significance. The court referred to *Shirlcar* (see p. 37) where Lawton LJ found such a heading rendered the notice doubtful and therefore invalid. However the Court of Appeal stated that any solicitor (read professional) reading the letter would have reasonably inferred that the phrase was meaningless.

This judgment, depressing as it would appear to encourage incompetence and uncertainty, should be noted but not followed.

It could now justifiably be asked whether there is any meaning whatsoever to be imputed to this phrase. Judgments of this tenor do nothing to produce clarity.

Dillon LJ, who gave the leading judgment, also participated in the *Shirlcar* decision. Norse LJ quoted the passage from *Shirlcar* of Dillon LJ, which concentrated upon the view that the heading would lead someone to view the contents as provisional. What is really the difference between the two situations?

'SUBJECT TO CONTRACT' – EFFECT ON COUNTER-NOTICE

Sheridan v Blaircourt Investments Ltd. (1984) 270 EG 1290

There were various items of correspondence from the tenant's solicitors, some of which were headed 'without prejudice and subject to contract'. The counter-notice, seeking a rent determined by a referee, as required by the lease, was made in a letter which was headed 'without prejudice and subject to contract'.

The judgment was for the landlord. The Court of Appeal decided that a reasonably sensible business man would not have realised that such a letter constituted a counter-notice.

'SUBJECT TO CONTRACT', TIME OF THE ESSENCE, AND A DEEMING PROVISION

British Rail Pension Trustee Co Ltd. v Cardshops Ltd. (1986) 282 EG 331

In a rent review provision the service of only one trigger notice by the landlord was permitted; in response to which a counter-notice was to be served by the tenant within eight weeks, of which time was to be of the essence. In a DEEMING provision, the lease stated that if the counter-notice was not served, then the rent specified in the trigger notice should be the 'market rent' and therefore the new rent for the period under review.

A trigger notice was served, various pieces of correspondence were exchanged, and, still within the eight-week period, the tenants issued a counter-notice. This counter-notice was made 'subject to contract'. The tenant considered that this notice was a valid counter-notice. The landlord disagreed.

The judgment was for the tenants. The court held that:

1. The test was would a 'reasonably sensible business man have been left in any doubt as to the effect and relevance of the counter-notice'? Would it be clear that the tenant intended to bring into operation the section in the lease relevant to agreement of a market rent and the events that should follow in the event of failure to agree?
2. A submission that it was important for a landlord to know not only to what figure the tenant would agree but also the figure which the tenant considered to be the market rent should be dismissed.
3. The 'subject to contract' phrase here does not actually trigger the rent review machinery, nor exercise a right to elect to have a rent determined by an expert or arbitrator, but merely triggers the period during which the parties are required to enter into negotiations to attempt to agree the rent. Therefore the proviso with the notice is sufficient to amount to an effective counter-notice.
4. The phrase 'subject to contract' is either present by mistake or designed to ensure the letter is not taken to be an offer capable of acceptance. This interpretation is consistent with the intent of a counter-notice.

There are several general points of interest in this case:

1. The 'notice' was not headed 'counter-notice'.
2. There was no indication that the offered counter rent was a market rent, yet the court deemed the notice sufficient.

3. The notice did not contain reference to either the need to request that the dispute be determined by an arbitrator or independent expert or the clause which requires agreement in good faith to be sought between the parties.

As a result, this is a somewhat surprising decision that would have been welcomed by the tenants. It is considered that, until it is tested by a higher court, it should not be relied upon. A more rigorous adherence to the conventions used in most counter-notices would be more cost-effective for all parties.

Where negotiations have failed to agree the value of the revised rent, the procedure should require the appointment of a third party to determine the revised rental level. On occasions this procedure will not be possible to implement.

Arbitrator or Expert where neither specified within Clause

It appears that where there is in the review clause a statement that the DETER-MINATION should be made subject to adherence to the Arbitration Acts of 1950, 1979, or 1996 then, in spite of indications that the determination should be by an independent expert, the determination should be made by an arbitrator. There is not at present any case reported upon this matter but this is the view of arbitrators and independent experts acting within this field.

In *Langham House Developments Ltd. v Brompton Securities Ltd.* (1980) 256 EG 719, the court decided that the determination should be made by an independent valuer and not an arbitrator for the reasons that:

1. There was no reference whatsoever to arbitration within the review clause in spite of the fact that reference to arbitration was made elsewhere within the lease.
2. The wording within the review clause pertained more to a determination by an independent expert than to that of an arbitrator.

ARBITRATION OR DETERMINATION – VALUATIONS AS ADMISSIBLE EVIDENCE

Seabridge Australia Pty Ltd. v JLW (NSW) Pty Ltd. and Others (4 April 1991) 101 ALR 147

The applicant accepted that rent review valuations were business records of the company. However it was argued that the valuations were not admissible EVIDENCE as the valuations were the conclusion of a judicial or administrative hearing. The rent review clause in question referred to the valuer as providing a DETERMINATION which shall be produced by the valuer acting as expert and not arbitrator. In the judge's view:

> Essentially an arbitration is an inquiry of a judicial kind that involves the determination of a dispute about existing rights according to an external (and therefore objectively ascertainable) standard or criterion. A valuation,

on the other hand, is an inquiry that presents a question to be answered on the basis of individual experience, expert knowledge and personal inquiry and investigation. The same or a similar question, such as market value may arise for determination in the course of both arbitration and valuation but the method of determining it and the legal consequences of doing so, are different.

The court ruled that the valuations were admissible.

DIFFERENCES BETWEEN EXPERT VALUATIONS AND AWARDS OF ARBITRATORS – A QUESTION OF IMPARTIALITY

Enron Australia Finance Pty Ltd. v Integral Energy Australia [2002] NSWSC 753

Here an expert known as a 'Reference Market-maker' was asked to fix a market quotation as between the parties. The court had to consider whether the appraisal was an award or an expert DETERMINATION. The key question was whether the decision must be made independently and impartially if the determination is to be made by an expert. The court considered that:

> It is of assistance to address this issue by first asking whether the Reference Market-maker's task is to be seen as that of an arbitrator, i.e., a quasi-judicial determination which will automatically invoke the principles of impartiality, or whether the task is merely that of an expert, valuer or appraiser.

The court cited with approval that:

> the existence of a dispute, although a factor, is not necessarily a decisive factor in determining whether arbitration or appraisement is involved. It is quite possible for parties to become involved in a dispute about something, such as the value of premises or goods, which they agree to submit for appraisal without intending that an arbitration should follow. The distinction depends upon a range of factors of varying importance and weight depending on the circumstances; but generally what must be in contemplation is that there will be 'an inquiry in the nature of a judicial inquiry'.

The court also noted with approval that: 'The arbitral function is to hear and resolve the parties' opposing contentions while an appraisal or expert decision is typically an appraisal in monetary terms of property value, or of loss or damage, made through specialist knowledge, or skills without any requirement of first hearing the parties.'

The learned judge went on to describe the role of the Reference Market-makers as being far removed from that of an arbitrator.

Their role is far from one which requires any input from the defaulting party and one which requires only a small amount of input from the non-defaulting party in terms of formal provision of necessary information to be taken into account in the exercise, such as the identification of the relevant swap contracts in respect of which the exercise is to be carried out.

Having decided that the role was that of an expert they had to decide the question of a need for impartiality. Here the submission was that the Reference Market-maker had acted in excess of his jurisdiction and was therefore impartial. The court found that there was a subtle distinction between the two actions. It would be possible to act outside the required jurisdiction without acting in any way impartially.

The court found that there was an implied term that an expert would have arrived at a valuation honestly and fairly.

The court determined that while there was an excess of jurisdiction there was no act of impartiality.

The Arbitrator

It is necessary to examine very briefly how the Arbitration Act of England and Wales 1996 affects rent review arbitrators. Up until 1996 arbitration was encompassed within the 1950 and the 1979 Arbitration Acts. The 1996 Act has not incorporated the United Nations Commission on International Trade Law (UNCITRAL) Model Arbitration Law much to the regret of some members of the original Marriott working group set up to draft a new Arbitration Act. Therefore the 1996 Act is a very strong but unique piece of legislation crafted solely for the purposes of English and Welsh arbitration.

Arbitration is always a second profession. Therefore in the specific area of rent review an arbitrator will always be either a chartered surveyor or a member of the legal profession. It is implicit in the appointment that the arbitrator will also be experienced in the giving of EVIDENCE in rent reviews (or advising, as in the instance of a legally qualified arbitrator) and from this expertise will have gained the respect of his colleagues and thus become an arbitrator. In addition to considerable practice experience, arbitrators will also (or should also) have what is known as a judicial capacity, i.e., be able to administer a hearing efficiently, effectively, and politely, while also respecting the rules of natural justice, s. 33.

The powers of the arbitrator are those which are given to him by the parties. He can only take unto himself the powers which both parties agree, or consent, to give him. Thus, if one party wants him to have power to proceed EX PARTE and the other side objects, then without further legal support the arbitrator cannot have that power. The Act appears to permit the parties to give that power but, it is suggested, that conclusion is not without an area for debate. It is not clear whether the power to proceed *ex parte* now exists, although s. 41 appears to authorise such action in certain circumstances.

Section 42 appears to nullify all the suggested powers of the arbitrator by stating explicitly that unless otherwise agreed by the parties a court may make an order requiring a party to comply with a peremptory order of the arbitrator. That implies that the arbitrator cannot take any peremptory action unless and until supported by the court.

There are various powers set out in the 1996 Arbitration Act. These are only available when there is an arbitration required to take place under the auspices of that Act. If the Act is specifically excluded from the terms of appointment then the powers of the arbitrator are completely at the discretion of the parties. Indeed a moot question is whether the Act can be excluded in whole or in part. There are Sections from which the parties cannot resile (derogate).

Section 34 of the Act makes clear that an arbitrator shall decide all procedural and evidential matters. However, and it is an important qualification, the parties have the right to agree any matter. This phrase is unclear. It appears to mean that the parties can agree ISSUES. In fact it is thought the meaning is that the parties can restrict the powers of the arbitrator as to all the matters set out in s. 34. The powers are very detailed unlike those of the UNCITRAL Model Arbitration Law. There is an obvious implication that where a power has not been identified then it is not available. Furthermore, there is no definition of what is a procedural matter and what is an evidential matter.

Without question all arbitrators will have to prepare extensive draft ORDERS FOR DIRECTIONS and in view of the risk of challenge the requirement to obtain agreement is very important. The vexed question arises as to whether an arbitrator can refuse to proceed if one party refuses to agree, while there is agreement from the other party. There does not appear to be any clear repetition of s. 5 of the 1979 Act which permitted arbitrators to seek the assistance of the court in certain circumstances, although s. 42 does approach that concept. It is thought that an arbitrator should not proceed without unity of purpose from the parties. This will not be easy to achieve. The arbitrator is the servant of the parties and must not forget that crucial point.

What is difficult is to relate s. 34 to s. 38(1). This latter section is an all encompassing authority given to the parties to agree upon the powers exercisable by the arbitrator for the purposes of and in relation to the proceedings. There is no reference to s. 34. Yet it is difficult to identify how the procedural matters identified in s. 34 are either separate from or related to the general context of those powers so particularly specified in s. 38.

In view of this apparent, or possible, divergence of power the advice to arbitrators is to be very clear and explicit in all 'procedural' matters. The arbitrator should preferably invite the parties to prepare a list of powers upon which he or she can, and must, comment if he or she considers that a power or rather an absence of a power is counter-productive.

This Act is a work of genius. When simplicity was required precision took its place. There will not be revisions for many years nor will they be required.

In the following cases, various distinctions are drawn between the powers of arbitrators and those of independent experts.

IS AN INDEPENDENT SURVEYOR APPOINTED BY THE PARTIES AN ARBITRATOR OR AN INDEPENDENT EXPERT?

North Eastern Co-operative Society Ltd. v Newcastle upon Tyne City Council (1986) 282 EG 1409

An independent surveyor, appointed by agreement, issued a REASONED AWARD having received WRITTEN REPRESENTATIONS from both sides. A tenant wished to challenge the award. Was the DETERMINATION carried out by a surveyor acting as an independent expert or as an arbitrator?·

The court decided that in this case the independent surveyor was an expert. The decision stated that: a mere reference to arbitration in the lease is not a clear indication, on its own, that an arbitration was intended to be the means of dispute settlement in every instance.

From the phrase 'In default of agreement as to independent surveyor – then by an arbitrator' it does not follow that the parties meant arbitral procedure to apply as they may have sought an independent surveyor from amongst local people whom both sides knew and respected, and who would follow the procedure of an independent expert.

EXPERT – JURISDICTION, EXCLUSIVE

National Grid Company plc v M25 Group Ltd. [1999] 1 EGLR 65

An expert was appointed to determine the review value. The tenants' proposal was £25,000 per annum. That of the landlord was £914,925 per annum.

The parties agreed that for the review various questions required answers. The issue was whether the court should answer the questions or the expert.

The Court of Appeal firmly found that the issue of jurisdiction went to the core of the role of the expert. Where there was any discretion then that discretion should be exercised in favour of the court as opposed to the expert.

The terms of the lease gave exclusive jurisdiction to the expert to determine the valuation. However the expert did not have jurisdiction as to the construction of the terms of the lease. The court would make those findings if so invited.

Many experts handle all aspects of the review with great competence. Some choose to have a LEGAL ASSESSOR sitting with them, others have sufficient experience and confidence to weigh the legal submissions without assistance. There is no need for a valuer to be legally qualified, what is required is judicial capacity and a flexible mind.

WHAT IS AN UMPIRE?

Safeway Food Stores Ltd. v Banderway Ltd. (1983) 267 EG 850

In the lease the following requirement existed: where both parties were required to appoint a valuer to agree the value of the rent, in the event of their failure to agree, then the President of the RICS should appoint an umpire. The parties disagreed as to whom to appoint as valuer. There was subsequently argument as to whether the umpire should be an expert or an arbitrator. The court decided that the umpire should be an independent expert because:

1. The matter of DETERMINATION of the market rent is one of expertise.
2. The requirement that an expert be a member of the RICS indicates that he must have expertise.
3. An express and separate reference existed in the lease to a separate arbitrator.

This was a somewhat surprising decision as the court implied that arbitrators are separate individuals from experts and also lack expertise. In reality, those members of the RICS who act as experts also can, and do, act as arbitrators. Indeed, it is not possible to be appointed to the panel of arbitrators of the RICS until expertise is acquired.

ORAL AGREEMENT TO ARBITRATE

Galliard Homes Ltd. v J. Jarvis & Sons plc 71 CON LR 219
A dispute had arisen between the parties. There was an attempt to enforce an arbitration agreement. The Court of Appeal found that there was an oral agreement made between the parties. However the parties were not agreed as to the terms of the agreement, which was made subject to contract. Therefore any arbitration agreement could not be incorporated into any notional contract.

APPLICATION FOR APPOINTMENT OF EXPERT

United Co-operatives Ltd. v Sun Alliance & London Assurance Co Ltd. (1986) 282 EG 91
In this case the tenants considered the landlord's application to the President of the RICS for the appointment of an expert to be premature. They objected, indicating that they were considering seeking a declaration as to the construction of the lease. The tenants sought an injunction against the landlord seeking to restrain it from taking any further step, and also attempted to join the President.

Hoffmann J. dismissed the application for an injunction by the tenants, and, in any event, the President of the RICS should not be a party to the application; he was not under any obligation to the parties not to make an appointment. If the tenants felt aggrieved they should have sought to restrain the expert.

APPLICATION FOR APPOINTMENT OF EXPERT – TIME OF THE ESSENCE

Banks and Another v Kokkinos and Another [1999] 2 EGLR 133
The review provisions required that a landlord should make an application to appoint an expert in the event that the tenant served an effective counter-

notice. The tenant did so. The landlord did not make an application for the appointment within the time limits. The landlord argued time was not of the essence.

The court found that the review provisions closely resembled those in *Vision-hire* (see p. 16). That case had identified a form of 'ultimatum procedure'.

The court found that there was an ultimatum procedure here. The landlord lost its right to a review.

This is another example of English courts giving persuasive authority to Scottish decisions, especially those which favour a stricter approach to time of the essence provisions.

MATTERS TO BE RESOLVED AT AN ARBITRATION – HOW SHOULD THEY BE DECIDED?

Burton v Timmis (1986) 281 EG 795

There was a dispute as to what occurred on a certain date, both as to what was said and as to what was agreed.

The Court of Appeal decided that because nothing was submitted in writing as EVIDENCE to the arbitrator it was not open to the arbitrator to resolve any conflict of evidence as to what was said or agreed.

As part of the *obiter dicta*, Kerr LJ stated that for an arbitrator to consider such matters when evidence had not been submitted must be an error of law on the part of the arbitrator but it is not necessarily misconduct.

It should be noted that this arbitration took place under the Agricultural Holdings Act 1948 and the facts appertaining to this case are peculiar to that Act.

WAS TIME OF THE ESSENCE FOR APPOINTMENT OF ARBITRATOR?

Staines Warehousing Co Ltd. v Montagu Executor & Trustee Co Ltd. (1987) 283 EG 458

Here both parties accepted that time was of the essence. A letter by the landlord's agents was sent to the President of the RICS within the time limits set down within the lease. Approximately six months later an application form and fee were sent to the RICS, this action taking place outside the time limits.

The Court of Appeal decided in favour of the landlord that time was not of the essence in this final action. Procedure for the RICS did not require a fee to be sent for there to be a valid application to appoint. There had been an application to appoint. Procedure merely meant the application would proceed no further until the fee was paid. Therefore a valid step was taken and valid application was made, albeit one called 'an in time only' application.

Cavoulakos and Others v 616550 Ontario Limited and Another 15 ACWS (3d) 368 18 May 1989, Ontario, Canada

The lease contained a provision to renew the lease. In that event there would be a reviewed rent. If there was a failure to agree then each party could appoint an arbitrator. The date for appointment was by 1 October 1988 with an award to be published within the twenty-one day period before the expiry of the lease. On 1 November the landlord's agent informed the tenant that arbitration was still available. There had been negotiations prior to that date. At that date the date for appointment had passed. Where one party had not appointed its arbitrator by the appropriate date the arbitrator appointed could fix the rent. The tenant had unbeknownst to the landlord appointed an arbitrator. As the landlord had not made an appointment the tenant's arbitrator fixed the rent.

There was evidence given as to faxed communications and non-receipt thereof.

The landlord appealed. The court found that time was not of the essence for the appointment of the party arbitrator in spite of clear and explicit wording or CONTRA-INDICATIONS.

The court found that to determine the value without hearing any evidence from the landlord was a breach of natural justice. The award was set aside.

Although the tenant acted in a very unusual way as to the actions of its arbitrator, prior to that event it had done everything it could to obtain co-operation from the landlord and had been ignored until after the deadline. The Ontario court took a very flexible view of the bargain between the parties. The circumstances before the court were very convoluted. A key aspect is the speed of the valuation.

Friesen v Martensville Gateway Plaza Service Centre Ltd. (7 August 1998) 173 SASK R 264

The lease had a term of six years. There was an all-encompassing time of the essence clause. There were discussions about a lease renewal. The expiry date of the original lease was 31 August 1998. The landlords issued an equivalent trigger notice for the rent for the proposed renewed term.

Six weeks later the agent of the tenant issued a counter-notice. This notice was out of time. That notice specifically sought the commencement of arbitration proceedings and set out a timetable for the proceedings. The landlord did not respond. Counsel for the tenant informed counsel for the landlord that the tenant was appointing its arbitrator. Under the time limits in the lease the landlord had three weeks to make its appointment.

The tenant applied for an order that the arbitration proceed. The court refused. Time was of the essence of every step and was an express CONTRA-INDICATION. The lease required that any nomination be made not less than ninety days before the expiration of the term. The tenant could have complied. It did not do so. The tenants were not children. The lease was required to expire at the end of the month.

Not only does this judgment indicate that effective drafting of a time of the essence clause can operate as required, it also indicates indirectly that Canadian arbitration has an element of alacrity sorely lacking in England and Wales.

TIME OF THE ESSENCE – DELAY IN APPLICATION FOR APPOINTMENT OF SURVEYOR

Iceland Food plc v Dangor and Others [2002] 21 EG 146

The lessor was required to serve a trigger notice not more than twelve months and not less than three months before the review date. Further, where the parties had failed to agree as to the revised rent then the market rent should be determined by a surveyor to be agreed upon in writing by the parties not less than one month before the due date. There was a default provision which directed that the trigger notice would be void if the landlord neglected to seek the appointment of an arbitrator within a specified time. The application had not been made. Nowhere was time described to be of the essence. However these directions could be described as clear CONTRA-INDICATIONS permitting an inference to be drawn to that effect.

A glance at the above procedure indicates that the timetable is almost impossible to achieve, but the parties have, at least, to try to achieve it. Alternatively, they should agree, on negotiated conditions, to change the terms of the timetable to accord with something more practicable.

The application had not been made within the time frame.

The court decided that in spite of the explicit timetable time was not of the essence. An application could still be made.

The court stated that to make time of the essence the tenant would have to serve an explicit notice to that effect upon the landlord.

There appeared to be a modicum of fudging the issue where the court stated that while the landlord was under a duty to proceed expeditiously with the review they had in fact acted within a reasonable time. The trigger notice was served almost four months after the expiry of the time limit and the application some four months thereafter.

These do not appear to be acts of expedition. While the provision was poorly drafted it does appear that the parties were not held to their bargain and the tenant suffered. Responsibility for the lease is notionally that of the landlord, it should be construed in favour of the tenant.

Extension of Time under s. 12 Arbitration Act 1996

The 1996 Act permits an application to be made to the court (or to the arbitrator where given such powers) to grant extensions of time to enable the arbitration procedure to continue. It will obviously only apply where there is a strict timetable and one party to the dispute has failed to adhere to it. The other party will, naturally, be unwilling to grant, by agreement, any extension of time. Therefore, the only recourse to the claimant party is to seek the assistance of the court or arbitrator. This power of the court is especially relevant when time is of the essence in all stages of the rent review clause. As will be seen it is very unlikely that an extension of time will be granted.

APPOINTMENT OF ARBITRATOR – EXTENSION OF TIME – ARBITRATION ACT

Fordgate (Bingley) Ltd. v *National Westminster Bank plc* (25 May 1995, Chancery Division, unreported)
The trigger notice could be served at any time over a twenty-four month period. The notice was effectively served. The application for the arbitrator had to be made within a specified time period. The application was made out of time by two months.

The court decided that the power to extend time for the commencement of arbitration proceedings was not just procedural but could have the effect of reinstating courses of action that would otherwise be lost.

The question was, should the discretion be exercised in favour of the landlord? The delay was two and a half months. The fault was the result of an oversight by the landlord's surveyor. The tenant was not prejudiced. The fact that the landlord could sue the negligent surveyor was of limited importance.

This case is noted almost as a reminder of the possible abuses of the arbitration legislation. It is very unlikely that an application made with the same facts will receive any sympathy at all under s. 12 of the Arbitration Act 1996, nor should it.

Fox & Widley v Guram & Another [1998] 1 EGLR 91
The judgment also makes important rulings upon trigger notices and counter-notices (see p. 29), which is the reason why the question of an extension of time being granted was reviewed.

This is a verbatim extract from the part of the judgment examining the extension of time aspect. It is crucial that all concerned are now aware how difficult it will be to succeed in this area. The court held that it had the power to grant an extension of time provided that the criteria established under s. 12 (1) were met:

> The question under subsection (3)(a) (of Section 12(1)) is whether the circumstances were such as were outside the reasonable contemplation of the parties when they agreed the provision in question and, if so, whether it would be just to extend the time.

The court agreed that, in considering this question, it is appropriate to take account of the underlying commercial purpose of provisions of the clause. Having taken into account all the circumstances, the court did not consider that these circumstances were such that they would not come within the reasonable contemplation of the parties.

Indeed it will be a very unusual situation that meets that criterion.

The court noted various exchanges of correspondence. There was a letter which was clearly without prejudice to the rights of the landlord. The court ruled that the tenant would not have been misled into thinking that it need not serve a counter-notice.

The tenant also relied upon s. 12(3)(b). It argued that the conduct of the landlords made it unjust to hold the other party to the strict terms of the contract. The same correspondence was relied upon to support the submission. A key point was that the landlord proposed a valuation figure which had little relationship to reality. The court determined that such behaviour was not unusual. It could not be said to cause injustice to a party (the tenant). Further the court had already ruled that the tenant had the opportunity to serve a counter-notice but had failed to do so. The court decided that there was no injustice within the meaning of the section. The application failed.

The grounds for an application under this section must be serious and of public importance and should not be valiant attempts to rescue a party from a situation in which it had failed to exercise due diligence.

EXTENSION OF TIME FOR COMMENCEMENT OF ARBITRATION UNDER THE ARBITRATION ACT 1996 – A SEA CHANGE

Harbour & General Works Ltd. v The Environment Agency [2000] 1 WLR 950
The regime for granting extensions of time for commencing arbitrations was reviewed by the Court of Appeal. The court below had refused to grant an extension. This judgment makes clear that the careful analysis of the judge in *Fox* (see case above) will epitomise the general problems faced where there is an attempt to obtain an extension of time.

There were procedures for resolving disputes. Within those procedures there were timetables. The appellant accepted that it was out of time when it served its notice seeking conciliation. The main contract had not placed a time limit upon that service but a time limit had been inserted in the corrigenda (additions or changes) to the contract.

The appeal rested upon which version bound the appellant, but the criteria were considered.

The court below stated that the former test of undue hardship had gone. In its place were the twin tests of circumstances beyond the reasonable contemplation of the parties at the time when the agreement was made and where the conduct of the respondent made it unjust to rely upon the time limit. The test should contemplate that at the time of the agreement the parties should consider that the time bar would not apply to them.

The Court of Appeal endorsed these views and refused LEAVE to appeal.

It appears that this dual test will make it very difficult for parties to obtain extension where time even appears to be of the essence. The attitude of 'tomorrow, tomorrow' just will not do. Section 12, the extension section, by implication expects every professional to do his duty by the parties they represent and keep in mind the *Basingstoke* (see p. 70) principles. If there is any doubt it is worth varying the lease even where it costs substantial funds. Money spent here could save many multiples in later reviews.

WHEN IS ARBITRATION APPOINTMENT EFFECTIVE? – DEFINITION OF A REFERENCE

University College Oxford v Durdy [1982] EGD 28

When is the appointment of an arbitrator effective? Was it the date when the letter of appointment was sealed and the subsequent acceptance of that appointment? The alternative was when the formal notification of the appointment was made to the parties.

The facts are not important save to state that a Minister made the appointment. The reason for the concern was that if time ran from the date of notification then it would be too late to comply with the time limits. However if time ran from the date of appointment then the appointment was in time.

The court held that an arbitration where the parties appointed the arbitrator was a commercial arbitration. The parties had to co-operate with one another in the appointment process. There was no reference whatsoever to the frequent instance of the failure to agree upon the appointee. In the case where a third party made the appointment that was a non-commercial arbitration. Thus the appointment was made upon appointment and not upon notification. As will soon be appreciated the majority of the appointments are made thus and are therefore non-commercial. The implications are of great concern.

The Court of Appeal added a very confusing comment to the effect that whenever the appointment was made it would be 'absurd' for time to start running until the parties were notified. That is not the point. With a trend that parties should be held to the time periods great injustice could emerge if this is not rectified. This judgment is said to be limited to the rare area of farm tenancy renewals. Those were the facts upon which the appeal was based. However the general comments and the misunderstanding of the commercial appointment process remains.

The late Sir Michael Kerr, as he then was, offered a definition of a REFERENCE as in 'a reference to arbitration'. It encompassed, at the latest, the appointment of the arbitrator and covered the period up to and including the final hearing and possibly the publication of the award. Whether this is the current view has to be considered.

Note: The Arbitration Act 1996 is silent upon this matter. There are three sections assisting the parties as to the procedure upon a failure to agree as to the appointment following closely those of the UNCITRAL Model Arbitration Law. The Model Law is also silent on this issue.

POWER AND SCOPE OF ARBITRATOR TO ORDER DISCOVERY

Cornwall Coast Country Club v Cardgrange Ltd. (1987) 282 EG 1664

The protection of privilege was claimed for the trading accounts of one of the parties to a lease. The party wishing to inspect the accounts sought the ruling of the court as to whether the qualification of privilege was justified.

The court held that such trading accounts may be admissible in certain circumstances. The following areas of guidance were decided by the court:

1. Admissible evidence may include the profit-earning capability of the property.
2. Where trading accounts are not statutory accounts or those accounts prepared to meet the requirements of various Companies or Partnerships Acts, then those trading accounts may be privileged, i.e., confidential. Those accounts should only be discovered when they are those which are generally available to the public.
3. The arbitrator will decide in each instance when the trading accounts not generally available would be available to the prospective lessee in a hypothetical open market.
4. It is for the arbitrator to decide what weight to give to any such evidence under this heading when it is admitted.
5. The arbitrator will hear submissions as to whether such trading accounts are in fact confidential, and will adjudge whether privilege attaches to them.

POWER OF ARBITRATOR TO INSERT TERMS IN LEASE TO REFLECT
CURRENT PRACTICE AS WELL AS TO DETERMINE A DIFFERENTIAL RENT

National Westminster Bank Ltd. v BSC Footwear Ltd. (1980) 257 EG 277
In this case there was a twenty-one-year lease with an option to renew for a further twenty-one years. At the option to renew, the rent could be determined by an arbitrator if it were not agreed by the parties. The landlord wished the renewed lease to be amended to take account of current review patterns and requested the arbitrator to order a differential rent. The tenant objected and the guidance of the court was sought.
    It was decided that:

1. The arbitrator was not empowered to incorporate current review patterns into the renewed lease unless express words existed in the former lease to permit such an action.
2. An arbitrator has no power to order a differential rent or a periodic rent with variations unless, of course, express words exist in the lease to the contrary.

REQUIREMENT AND POWER OF ARBITRATOR TO DECIDE POINTS OF LAW

L.S. Sethia Liners Ltd. v State Trading Corporation of India [1985] 1 WLR 1398
In this case there was a POINT OF LAW for decision. One of the parties considered that arbitration was not the correct formula for such a decision: only the courts could decide such matters.
    The Court of Appeal decided that the arbitrator should decide the point of law. Although the matter before the Court of Appeal arose from the Arbitration Act 1979 the judgment of Kerr LJ pertains to all arbitrations:

Under the 1979 Act in the decision of the House of Lords in *Pioneer Shipping Ltd. v B.T.P. Tioxide Ltd. (The Nema)* and *Antaios Compania Naviera SA v Salen Rederierna AB*, it was only in a very limited number of cases that the court was given the opportunity to determine ISSUES of law in disputes which parties had agreed to refer to arbitration.

This strict 'hands off' regime continues under the Arbitration Act 1996.

ROLE OF THE COURT VIS-À-VIS AN EXPERT AND FINDINGS ON ISSUES OF LAW

PosTel Properties Ltd. and Another v Greenwell and Another [1992] 47 EG 106
In a lease where the DETERMINATION of value was to be made by an independent expert the question arose as to whether issues of law could be decided by the court before the determination had been made.

The judge concluded that once the determination had been made it could not be challenged. He relied upon the Court of Appeal decision in *Jones v Sherwood* (see p. 90). Therefore it was necessary for the parties to provide all the assistance possible for the expert before determination was made.

In this case there was a point of construction as to the assumptions to be made when analysing a USER CLAUSE. The court felt justified in giving guidance to the expert in this matter in order to expedite the expert's determination.

Power of Arbitrator to Issue Subpoena for Persons or Documents

The arbitrator has no power to issue or order a subpoena for either documents (i.e. comparable EVIDENCE) or a person.

Costs

Costs decided by the Parties

Parties may, by agreement, decide how the costs shall be divided, after the dispute has arisen.

Costs decided by the Arbitrator or Expert

The arbitrator has a discretion as to his award of costs. There is no necessity, as is sometimes believed, that in all rent reviews each party bears his own costs. Thus, where one party has put forward inappropriate and therefore time-wasting EVIDENCE of comparables, the arbitrator should be encouraged to use his discretion and take that fact into consideration when awarding costs. There are many other instances when such discretion should be exercised and it might well be that more rigorous use of such discretion might encourage parties to come to a settlement more speedily.

It is not clear how much discretion the independent expert has as to an award of costs. However, there appears to be no reason why at the preliminary meeting the parties cannot agree to give the independent expert such a discretion.

ABILITY OF ARBITRATOR TO CORRECT 'AN ERROR ARISING FROM AN ACCIDENTAL SLIP'

Mutual Shipping Corporation v Bayshore Shipping Co, The Times, 14 January 1985

Here the arbitrator, in his reasons, attributed the EVIDENCE of an expert to the wrong party. He accepted the evidence of that witness, and relied upon it in his award.

The Court of Appeal decided that such an error could be corrected once the arbitrator became aware of it. The Master of the Rolls said that 'Section 17 [Arbitration Act 1950] was directed to clerical mistakes in the award, which the present case was not, and to errors in the award, which it was.' This was an error arising from an accidental slip.

Earlier the judgment stated that where arbitrators become aware of such mistakes (which are not second thoughts) they should apply to the parties, or to the court if need be, to have the award remitted to them for amendment.

MISTAKES AND OMISSIONS FROM AWARDS – ARBITRATION ACT 1996

Bristol Myers Squibb Co v Baker Norton Pharmaceutical Inc [2001] EWCA 414 (Court of Appeal, unreported)

The issue in this appeal was the extent and occasion when an order of a judicial tribunal can be corrected. This is applicable to an arbitrator but not to an expert.

The argument was that a mistake as to the legal effect was not a slip and could not be corrected.

The Court of Appeal noted the *Bayshore* judgment (see case above). That concerned the misapplication of names to parties. The tribunal cannot have second thoughts. If the EVIDENCE has been wrongly assessed or the law mis-construed then the result cannot be altered. Goff LJ (as he was then) refused to define a slip but suggested that it was usually recognisable.

Here the terms of the order did not represent the judgment. It could be changed. The impact of the judgment appears very restrictive as to the scope of amendment of any interim or final award.

However s. 57 of the Arbitration Act 1996 permits an arbitrator, on his own initiative or at the request of a party, not only to correct slips and omissions but also to clarify or remove any ambiguity. So in a sense the Act permits an arbitrator to have second thoughts under the heading of clarification. This scope of review does not appear in the UNCITRAL Model Arbitration Law, Article 33, where any request for interpretation comes from the parties. It is not a good idea to permit an arbitrator to make these clarifications by reason of the width of opportunity.

Evidence and the Expert Witness

The major case, which sets out both how the expert acquires his knowledge and what is the best form of EVIDENCE, is *English Exporters (London) Ltd. v Eldonwall Ltd.* [1973] 1 Ch 415. The judgment is therefore quoted at length:

> Let me ... confine myself to the ADMISSIBILITY of hearsay in chief and in re-examination in these valuation cases. In such circumstances, two of the heads under which the valuer's evidence may be ranged are opinion evidence and factual evidence. As an expert witness, the valuer is entitled to express his opinion on matters within his field of competence. In building up his opinions about values, he will no doubt have learned much from transactions in which he has himself been engaged, and of which he could give first-hand evidence. But he will also have learned much from many other sources, including much of which he could give no first-hand evidence. Textbooks, journals, reports of auctions and other dealings, and information obtained from his professional brethren and others, ... have contributed their share. Doubtless much, or most, of this will be accurate, though some will not, and even what is accurate so far as it goes may be incomplete, in that nothing may have been said of some special element which affects values. Nevertheless, the opinion that the expert expresses is none the worse because it is in part derived from matters of which he could give no direct evidence. Even if some of the extraneous information which he acquires in this way is inaccurate or incomplete, the errors and omissions will often tend to cancel each other out; and the valuer, after all, is an expert in this field, so that the less reliable the knowledge that he has about the details of some reported transaction, the more his experience will tell him that he should be ready to make some discount from the weight that he gives it in contributing to his overall sense of values. Some aberrant transactions may stand so far out of line that he will give them little or no weight. No question of giving HEARSAY EVIDENCE in such cases; the witness states his opinion from his general experience.
>
> On the other hand, quite apart from merely expressing his opinion, the expert often is able to give factual evidence as well. If he has first-hand knowledge of a transaction, he can speak of that. He may himself have measured the premises and conducted the negotiations which led to a letting of them at £x, which comes to £y per square foot; and he himself may have read the lease and seen that it contained no provisions, other than some particular clause, which would have any material effect on the valuation; and then he may express his opinion on the value.
>
> So far as the expert gives factual evidence, he is doing what any other witness of fact may do, namely, speaking of that which he has perceived for himself. No doubt in many valuation cases the requirement of first-hand evidence is not pressed to an extreme: if the witness has not himself measured the premises, but it has been done by his assistant under his supervision, the expert's figures are often accepted without requiring the

assistant to be called to give evidence. Again, it may be that it would be possible for a valuer to fill a gap in his first-hand knowledge of a transaction by some method such as stating in his evidence that he has made diligent enquiries of some person who took part in the transaction in question, but despite receiving full answers to his enquiries, he discovered nothing which suggested to him that the transaction had any unusual features which would affect the value as a comparable. But basically, the expert's factual evidence on matters of fact is in the same position as the factual evidence of any other witness.

Further, factual evidence that he cannot give himself is sometimes adduced in some other way, as by the testimony of some other witness who was himself concerned in the transaction in question, or by proving some document which carried the transaction through, or recorded it; and to the transaction thus established, like the transactions which the expert himself has proved, the expert may apply his experience and opinions, as tending to support or qualify his views.

That being so, it seems to me quite another matter when it is asserted that a valuer may give factual evidence of transactions of which he has no direct knowledge, whether per se or whether in the guise of giving reasons for his opinion as to value. It is one thing to say 'From my general experience of recent transactions comparable to this one, I think the proper rent should be £x': it is another to say 'Because I have been told by someone else that the premises next door have an area of x square feet and were recently let on such-and-such terms for £y a year, I say the rent of these premises should be £z a year.' What he has been told about the premises next door may be inaccurate or misleading as to the area, the rent, the terms and much else besides. It makes it no better when the witness expresses his confidence in the reliability of his source of information: a transparently honest and careful witness cannot make information reliable if, instead of speaking of what he has seen and heard for himself, he is merely relating what others have told him. The other party to the litigation is entitled to have a witness whom he can cross-examine on oath as to the reliability of the facts deposed to, and not merely as to the witness's opinion as to the reliability of information which was given to him not on oath, and possibly in circumstances tending to inaccuracies and slips. Further, it is often difficult enough for the courts to ascertain the true facts from witnesses giving direct evidence, without the added complication of attempts to evaluate a witness's opinion of the reliability, care and thoroughness of some informant who has supplied the witness with the facts that he is seeking to recount.

It therefore seems to me that details of comparable transactions upon which a valuer intends to rely in his evidence must, if they are to be put before the court, be confined to those details which have been, or will be, proved admissible evidence, given either by the valuer himself or in some other way. I know of no special rule giving expert valuation witnesses the right to give hearsay evidence of facts: and ... I can see no compelling reasons of policy why they should be able to do this. Of course, the long-established technique in adducing expert evidence of asking hypothetical questions may also be employed for valuers. It would, I think, be perfectly proper to ask a valuer 'If, in May 1972, no. 3, with an area of 2,000 square feet, was let for £10,000 a year for seven years on a full repairing lease with no unusual terms, what rent would be appropriate for the premises in dispute?' But I cannot see that it would do much good unless the facts of the hypothesis are established by admissible evidence; and the valuer's statement that someone reputable had told him these facts, or that he had seen them in a reputable periodical, would not in my judgment constitute admissible evidence....

Putting matters shortly, ... in my judgment a valuer giving expert evidence in chief (or in re-examination):

(a) may express the opinions that he has formed as to values even though substantial contributions to the formation of those opinions have been made by matters of which he has no first-hand knowledge;
(b) may give evidence as to the details of any transactions within his personal knowledge, in order to establish them as matters of fact; and
(c) may express his opinion as to the significance of any transactions which are or will be proved by admissible evidence (whether or not given by him) in relation to the valuation with which he is concerned: but
(d) may not give hearsay evidence stating the details of any transactions not within his personal knowledge in order to establish them as matters of fact.

To those propositions I would add that for counsel to put in a list of comparables ought to amount to a warranty by him of his intentions to tender admissible evidence of all that is shown on the list.

THE ROLE OF AN EXPERT WITNESS

National Compania Naviera SA v Prudential Assurance Co. Ltd. (The Ikarian Reefer) [1993] 2 EGLR 183

This is the key aspect of the judgment referring to the duties of an expert witness in whatever field.

The duties and responsibilities of expert witnesses in civil cases include the following:

1. Expert EVIDENCE presented to the court should be, and should be seen to be, the independent product of the expert uninfluenced as to form or content by the exigencies of litigation: *Whitehouse v Jordan* [1981] 1 WLR 246 at 256, *per* Lord Wilberforce.
2. An expert witness should provide independent assistance to the court by way of objective, unbiased opinion in relation to matters within his

expertise: *Polivitte Ltd. v Commercial Union Assurance Co plc* [1987] 1 Lloyd's Rep 379 at 386, Garland J and *Re J* [1990] FCR 193, Cazalet J. An expert witness in the High Court should never assume the role of an advocate.

3. An expert witness should state the facts or assumptions upon which his opinion is based. He should not omit to consider material facts which could detract from his concluded opinion (*Re J, supra*).

4. An expert witness should make it clear when a particular question or issue falls outside his expertise.

5. If an expert's opinion is not properly researched because he considers that insufficient data is available, then this must be stated with an indication that the opinion is no more than a provisional one (*Re J, supra*). In cases where an expert witness, who has prepared a report, could not assert that the report contained the truth, the whole truth and nothing but the truth without some qualification, that qualification should be stated in the report: *Derby & Co Ltd. and others v Weldon and others, The Times*, 9 November 1990, *per* Staughton LJ.

6. If, after exchange of reports, an expert witness changes his view on a material matter having read the other side's expert's report or for any other reason, such change of view should be communicated (through legal representatives) to the other side without delay and when appropriate to the court.

7. Where expert evidence refers to photographs, plans, calculations, analyses, measurements, survey reports or other similar documents, these must be provided to the opposite party at the same time as the exchange of reports (see *Guide to Commercial Court Practice, 15.5*).

ADMISSIBILITY OF EVIDENCE

Rogers v Rosedimond Investments (Blakes Market) Ltd. (1978) 247 EG 467

Here the judge heard various parties give EVIDENCE. Amongst the parties were two traders who gave evidence for the tenant as to the correct level of rent for the property; they were not valuers. Valuers gave evidence for both sides as to the correct level of rent. The judge based his valuation upon the evidence of the traders – by implication ignoring the experts.

The landlord challenged the valuation. The Court of Appeal agreed that the judge should not have ruled upon the evidence of amateurs whilst ignoring the experts. The evidence was inadmissible. The court also confirmed that evidence given by experts who had no first-hand knowledge of the transaction was hearsay and therefore also inadmissible.

FURTHER GUIDANCE

Showan and Another v Yapp and Another (3 November 1998, Court of Appeal, unreported)

A cottage was subjected to occasional flooding. The plaintiffs bought the cottage. The plaintiffs discovered the extent of the flooding. They sued the former vendor for misrepresenting the extent and amount of the flooding. The amount of damages would be the diminution in value. The court below awarded damages. The plaintiffs challenged the valuation EVIDENCE upon the judge made the award. The defendant's valuer awarded £12,000 whereas the valuer for the plaintiff estimated a diminution of £90,000. The award was £12,000.

A lay witness supported the evidence of the defendant's expert. That evidence should not have been given any weight. Further there was additional expert evidence for which the proper evidentiary foundation had not been laid nor was it tested in cross-examination. The transcript of the evidence of the defendant's expert did not impress the Court of Appeal as to its certainty. Also it was based upon instructions which did not relate to flooding, merely water penetration.

The appeal was allowed. By consent the amount awarded was raised to £40,000.

There was no justification for choosing the figure of £40,000. The issue should have been adjourned to allow the expert evidence to be assessed by an expert or at least the proper forum. If the plaintiff's expert figure was correct and the cost of works to protect the property in such circumstances are significant then the plaintiffs continued to suffer financial loss in spite of victory.

HEARSAY EVIDENCE

Town Centre Securities Ltd. v Wm Morrison Supermarkets Ltd. (1981) 263 EG 435

Here the arbitrator listened to EVIDENCE given by an expert who had no direct knowledge of it: it was therefore HEARSAY EVIDENCE. No objections were raised at the time to the fact that the evidence was inadmissible. However, there was an appeal by the side who did not give hearsay evidence against the fact that the arbitrator apparently raised no objection to the evidence.

The judge decided that unless the evidence was objected to at the time when it was given, then for the sake of convenience such evidence should be treated as admissible.

Living Waters Christian Centres Ltd. v Fetherstonhaugh [1999] 2 EGLR 1

The court below had refused to issue directions to an arbitrator. There was also a refusal to set aside an award on the alleged ground that the arbitrator had given weight to 'unacceptable' comparables. There were very clear directions from the arbitrator as to the way that he would accept EVIDENCE, whether first hand or hearsay. All evidence had to be adduced in the form of a sworn affidavit. The tenant produced a report, which relied somewhat upon HEARSAY EVIDENCE. The report of the landlord also relied upon hearsay evidence. The evidence was not presented in affidavit form.

The tenant's representative complained to the arbitrator as to the form of the landlord's evidence. The arbitrator responded that any question as to ADMISSIBILITY was for him to decide. He also invited the parties to address him upon the issue of admissibility. The landlord submitted that it was not clear whether the tenant disputed the truth of the alleged hearsay evidence. That comment was not resolved.

The arbitrator stated in the award that the comparables were very unsatisfactory. Only one witness had produced a valuation in a standard format.

The court found that the arbitrator had done his best with limited evidence. The Court of Appeal agreed and dismissed the appeal.

This is another example of an arbitrator with the best will in the world not enforcing his own directions. If they are not to be enforced the arbitrator should not make the order. The direction as to the format was reasonable and rational. All submissions should have been rejected until they complied with the directions. This refusal will cause annoyance but is the only correct method to avoid the risk of a challenge.

AWARD OF ARBITRATOR SOUGHT TO BE PUT IN AS EVIDENCE

Land Securities plc v Westminster City Council [1992] 2 EGLR 13
The question came up which must arise frequently but is seldom litigated. The landlord wished to put in as EVIDENCE the award of an arbitrator determining the rent of an adjacent building. Was such an award admissible in evidence?

The court said it was not admissible. The arbitrator of the adjacent building had made his finding based upon evidence submitted to him and not upon his own expertise. Therefore, the present arbitrator would not have had any first-hand knowledge of any of the transactions submitted to him. It would be wrong to admit the award as evidence.

However, the court might have come to a different view had they been considering the DETERMINATION of an expert. For in that case the expert while noting the submissions made to him would be relying upon his own knowledge. In that case the determination might be admissible. Note the interesting case of *Moore Stephens* discussed below (see p. 114), not quite to the point but relevant.

EVIDENCE WHICH MAY BE USED AND ASSUMPTIONS WHICH MAY BE MADE

Ritz Hotel (London) Ltd. v Ritz Casino Ltd. [1989] 2 EGLR 135
Various preliminary points of law were raised for DETERMINATION by the court in order to assist the arbitrator.

The review clause was set out as follows:

... taking no account of (i) any effect on the rent of the fact that the tenant or any company with the same group ... has been in occupation of the demised premises (ii) any effect on the rent of any improvements ... (iii) any GOODWILL attached to the demised premises by reason of the business carried on thereat ... and it is expressly agreed that in assessing the market rent no account shall be taken of the turnover or profits of the business carried on by the tenant in the demised premises and that any Arbitrator ... shall not be entitled to call for or inspect the accounts of the business.... But having regard in so far as possible to the rental values then current for similar properties let on similar terms with VACANT POSSESSION for a term equivalent to the term hereby granted without a PREMIUM with the same provision for rent review and ... on the basis that ... the tenant does not hold but will immediately be able to obtain a licence under the Gaming Act 1968.

Was the hypothetical term twenty-one years (the original term) or the unexpired term? It was decided that the phrase 'term equivalent to the term hereby granted' meant twenty-one years from the date of grant. Therefore, at each review period there was a lesser actual unexpired term – this unexpired term became the hypothetical term. Indeed, the learned judge said, 'It is easy to imagine circumstances ... where the assumption on ... the fifteenth year of the term of a fresh letting for twenty-one years would produce a manifestly unjust result.' It is strange, but surely the imposition of the assumption of a hypothetical fresh term at each review in order to give a consistent rental growth pattern was what was intended, however unjust?

The learned judge also reminded the parties that the fact that eventually there would be a review with only one year unexpired would not have a disproportionate result. This rent reviewed for one year would be very useful whilst the landlord was awaiting the determination of the interim rent! It could be very unfortunate if this line of reasoning is followed. The purpose of the interim rent is to soften the blow for tenants of the new rent. The interim rental level for a new lease of fourteen years is likely to be higher than the rent of an unexpired term of one year with a possible statutory right to renew. Therefore, to indicate that the one-year review would be useful for renewal purposes is an argument that a landlord might not find unattractive.

The reason for drafting a lease where there was a review for one year only is unclear; such a pattern is not advisable. If the current review pattern is three years then grant a twenty-one-year lease, but if four or five years then a twenty-year term will be sufficient.

The next question concerned the ability of the arbitrator to look outside any direct comparable EVIDENCE, in fact to profitability. The tenant pleaded that outside evidence could only be examined if there were no direct evidence available. The learned judge considered that 'in the real world it is very unlikely that there will be direct evidence of near contemporaneous lettings of comparable properties. It is for the arbitrator to evaluate such evidence as there may be of lettings of other properties in the light of any other evidence relevant to the determinations of rental values which the parties may wish to adduce.' In other words, the arbitrator could use profitability in order to evaluate what evidence there was. The learned judge also decided that the arbitrator could take into account the profits and turnover of the group of companies to which the tenant belonged. The precise turnover and profit of the particular tenant must be removed from the arbitrator's mind, while the requirement that any goodwill must be disregarded meant that the tenant would be assumed to be 'starting from scratch' in setting up the business.

The next point to determine was whether the actual gaming licence was deemed to have been taken away or not. The decision was that, 'the plain purpose of the provision is notionally to take away the licence held by "the tenant" and to provide that the market rent be ascertained on the footing that any person minded to take an interest in the premises would be able to make his offer in the knowledge that he would be granted a licence contemporaneously with the grant of the lease.'

Finally the prospective tenant was deemed to be a tenant who was unlikely to have the licence cancelled having been awarded it.

This is a judgment which requires careful assessment.

Comparable Evidence

This means the rental EVIDENCE obtained from various other premises by the professional valuers on behalf of both the landlord and the tenant, and it will be put forward to support the value which both parties seek to defend.

The major and almost insuperable problem for both the opposing parties and the valuer is to be able to obtain such first-hand evidence of all the relevant facts pertaining to the comparable evidence as to be able to make a true value judgement as to the relevance of the rental value pertaining thereto.

It is generally accepted that rent obtained upon a review never reflects true market rental value, the reason being that although the presence of the tenant is hypothetically discounted, the landlord will be unlikely to be able to obtain full market rental value unless, of course, he insist on arbitration. The very presence of the tenant indicates that the rental value will be a result of a compromise between the parties, and this will ensure that the maximum rent is not obtained.

The best evidence to be used as comparable is that of a fresh open market letting. If this were not so it would not so often be the case that a landlord will buy in a lease when the opportunity arises in order to relet the premises on an open market basis thus providing himself with indisputable current market evidence for any other properties he owns in the area.

The written presentation of comparable evidence does not in itself reveal the effectiveness of the professionals on either side in their negotiations, or the financial position of the landlord, or any subtle relationship which may exist between the landlord and the tenant, e.g. where the tenant may be a subsidiary company of the landlord. Another problem with comparable evidence is that it is generally accepted by the profession that it is not necessary for professional valuers to give first-hand evidence of all aspects of the comparable evidence. This again must reduce the value of the evidence that is given.

One area of importance in determining the appropriateness of comparable evidence is the effect of the payment of any premium on an existing rental level. There are arguments on either side as to whether and how evidence of premiums as to value should be taken into account.

Premiums

A PREMIUM is a capital sum paid by an ingoing tenant for the right to occupy/ possess leasehold premises. There are other definitions which do not concern this element of valuation. In addition the tenant will also be paying the rent which applies on an annual basis to the premises as set out in the lease. There is some debate as to the relevance of a premium in attempting to determine the current market value. Valuers for a tenant at a rent review, when premium evidence is put forward by a landlord, will attempt to argue that the premium evidence has no relevance whatsoever as it was merely paid for the GOODWILL which existed to the premises and also it accounts for the value of fixtures and fittings left by the previous tenant.

This is surely right where a tenant takes over the premises, where the business which he intends to operate is the same as the previous business operated in that premises and the new tenant hopes both to retain the goodwill of the existing customers and to utilise effectively the fixtures and fittings of the previous tenant. However, this argument must be spurious when a new tenant takes over an empty premises devoid of fixtures and fittings and intends to operate a business of an entirely different use from that of the previous tenant.

At this point the valuers for the tenant will argue that the premium paid did not reflect the fact that the market rent for the premises was lower than current full market rental value, but was merely a freak factor which expressed itself by the tenant's idiosyncratic desire to be in that very position or in that very premises. This, it is argued, in no way reflects what the current market value of the premises is, it merely reflects what some unusual tenant would pay to enter the premises.

However, is not a willing tenant someone who is prepared to pay what he feels is full market rental value? If the tenant feels that full market rental value is the current rent together with the premium, then that surely is the full market rental value in the eyes of the trading public in general. Surprisingly, this is not the view of the profession as a whole, and the obvious evidence that a premium indicates of the rental level being lower than its true level is discounted.

Possibly the only way for a landlord to ensure that a premium paid by an assignee for the right to enter premises to some degree reflects partial or whole rental level is for the landlord, when the licence to assign is granted by it, also to require that the premium paid is broken down into various effective areas. This needs to be subsequently certified by the tenant as true. Thus, in an instance where an outgoing tenant receives a premium of £20,000 from the proposed assignee for a lease under which the user is completely different, and for premises which are empty, a statement is made by the assignee that the premium paid reflected, together with that rent currently payable, the total sums which he would be prepared to pay to enter the premises. That EVIDENCE, attached to the lease, as it would be, would surely be additional effective evidence of the true current market value of the premises at a later rent review.

Discounts

These are either features in the actual building or clauses in the lease, or both, which result in the demised premises (building) being worth less than the full open market rental value. A feature might consist of a very restricted frontage, restricted internal areas (listed buildings), or columns which by their very presence inhibit the clear operating space of the structure.

A clause in the lease which might have the effect of discounting the rent might be, for instance, a restriction upon the use of the premises. This might vary from the use restricted to a specific tenant to a use within a class or category. Any restriction upon the subsequent use of the premises upon assignment or other parting of possession of the property by its very nature limits the possible number of assignees, therefore the number of people available to take the premises.

GUIDANCE AS TO VALUATION

Cumshaw Ltd. v Bowen [1987] 1 EGLR 30
Here there was an originating summons as to whether the Retail Prices Index still existed in the form envisaged in June 1960.

The decision of the court was that the current index was the same – all that was needed was the application of a relatively simple formula in order to achieve an accurate use of the original mathematical base.

VALUATION – THE EFFECT OF A WIDER THAN NORMAL REPAIR CLAUSE

Norwich Union Life Insurance Society v British Railways Board (1987) 283 EG 846
In this case there was a hundred-and-fifty-year lease and a repairing clause which stated that the tenant should keep 'the demised premises in good and substantial repair and condition and when necessary to rebuild, reconstruct and replace the same and in such repair and condition to yield up ...'.

The arbitrator decided that this phraseology meant more than mere repair. Therefore the duty was more onerous than usual for the tenant. It was not unforeseeable that the building might outlive its natural life over a period of a hundred and fifty years. The arbitrator incorporated a discount of 27.5 per cent of full market rental value; this took into account various matters amongst which was the effect of this clause.

The landlord appealed on the ground that the arbitrator was wrong in law in giving such a wide meaning to the phrase in question. The court supported the arbitrator, and agreed that such a clause was more onerous than the normal repairing clause.

Before the decision can be taken as to the new rental value, it is vital to determine exactly what is the date upon which the new value shall apply.

VALUATION DATE – HOW DEFINED

Parkside Knightsbridge Ltd. v The German Food Centre Ltd. (1987) 18 CSW 53

In this case a deed of amendment to a lease described part of the rent review procedure as follows:

> The landlord shall be entitled to require by notice in writing to the tenant given on or before 24th June 1985 that the rent hereby reserved and payable hereunder shall be reviewed on the basis of the then exclusive rack rent value of the demised premises, so that improvements carried out by the tenant at its own cost with the consent of the landlord shall be ignored in ascertaining such rack rent value. Such yearly rent shall be from the 25th day of December 1985 increased....

The court decided that the use of the word 'then' in the second element of the rent review clause indicated that 24 June 1985 was the date at which the new rent should be determined.

MEANING OF 'AT THE DATE OF REVIEW'

Prudential Assurance Co Ltd. v Gray (1987) 283 EG 648

In this lease there was a clause which stated 'for the purpose of this clause the expression "commercial yearly rent" shall mean the yearly rent at which the demised premises might reasonably be expected at the date of review to be let in the open market by a willing landlord for a term of 14 years under a lease on the same terms and conditions in all other respects as this present lease there being disregarded.'

The date from which the reviewed rent was payable was 24 June 1986. The landlord indicated that the date upon which the valuation was actually determined, whether by agreement or by arbitration whenever it should take place, was the date of review. The tenant contended that the date for valuation should be as at 24 June 1986.

The landlord relied upon the phrase 'at the commencement of the term', three times repeated in the previous clause and referring specifically to 24 June 1986, to indicate that the draftsman intended that that date and the review date should not necessarily coincide.

The learned judge agreed with the contention for the landlord.

ASSUMPTIONS AS TO THE DATE OF VALUATION

Kaley Ltd. v Hong Kong Land Co Ltd. [1990] 1 HKC 381 and [1990] 2 HKC 194

There was a rent review provision, which required the appointment of an expert in the event of a disagreement as to the revised value. There was a disagreement. The parties also disagreed as to whether the expert should adopt any particular date and if so which date. The assumptions included the hypothesis that the premises were being let vacant for a term equal to the whole of the said term without reference to any other adjustments of rent but otherwise having note of the various disregards.

Where the formula was silent as to the date of valuation then the date should not be the date of assessment but the date at which the review was scheduled to start. That conclusion ensured that the tenant was not placed at a disadvantage by the inaction of a landlord where time was not of the essence.

The Court of Appeal supported the decision of the lower court and firmly rejected the appeal.

EFFECTIVE DATE FOR REVIEW

Ladbroke Group plc v Bristol City Council [1988] 23 EG 125

Initially, there was a building agreement between the parties. This was converted into a hundred-and-twenty-five-year lease from 5 July 1973. There were provisions for rent reviews at various times. The date of the first review was to be delayed until either a road was constructed or the landlord certified that the road was not to be built.

The road was not built on 4 July 1980. Therefore, the first review period was delayed. Later the landlords informed the tenants that the road would not be built. Subsequently, they considered that the next review date would be 5 July 1981. The tenants, however, considered that the delay commenced from the date of the start of the lease, that is 1973. Therefore, the delay with relevance to the road was not one year, but seven years.

The Court of Appeal considered the draft lease. The draft made it clear that the date to which the delay applied was 31 March 1973. Therefore, since the landlord did not serve notice on the tenants for seven years plus an additional part of a year, the review would be delayed for that period.

DATE OF VALUATION FOR RENT REVIEW PURPOSES – DETERMINATION BY SURVEYOR OF COMMENCEMENT OF REVIEW PERIOD

Glofield Properties Ltd. v Morley and Another (no. 2) [1988] 2 EGLR 149

The parties disagreed as to the date to which the rental value should apply. The tenants considered that the value for the new review period should be the current market value as at the date upon which the new review period commenced. The landlord, however, considered the value should be that at the date upon which the independent surveyor made his award.

The tenant protested. It considered that the meaning sought by the landlord could only benefit the landlord. As the landlord was in control of the triggering process, it could delay the process and benefit from the delay by reason of the rise in rental values. The clause stated:

> the expression 'the open market rental value' as aforesaid means a sum in relation to the review period or the second review period as the case may be determined in manner hereinafter provided as being at the time of such determination the annual rental value of the Demised Premises in the open market on a lease for a term of years being equivalent to the number of years then unexpired of the term granted by this lease with VACANT POSSESSION at the commencement of such term.

The judge indicated that the meaning sought by the landlord did appear unfair. It did benefit the landlord should a delay occur between the time when the tenant sought a determination by the independent surveyor and the actual date of the determination. However, where there was a clear statement of the desires of the parties, that should not be ignored unless within 'the four corners' of the lease there were evidence to the contrary.

Various Definitions of Rental

For some strange reason those who draft leases and licences vary in the definition of the rent return sought by the landlord. An analysis of the decisions of the courts as to the actual meanings of the various lease phrases indicates that there is very little difference between any of them. However, it is sensible at least to be aware of any notional differences.

Open Market Rack Rent   This basically means the demised premises vacant and to let by a willing landlord to a willing tenant, with any necessary assumptions or disregards taken into account.

Open market rent   In *St Martins Property Investments Ltd. v CIB Properties Ltd. and Another* (11 November 1998, Court of Appeal, unreported), the review rent was to be the open market rent, which was defined as:

> The full yearly open market rent for which the demised premises might reasonably be expected to be let as a whole on the relevant review date in the open market by a willing landlord to a willing tenant with VACANT POSSESSION without taking a PREMIUM.
> 1.5.2 For a term equal in duration to the original term hereby granted and on the basis (whether or not it is a fact) that the demised premises enjoyed planning permission for the user of the demised premises for the time being authorised by this Underlease for the same period.
> 1.5.3 Otherwise upon the terms and conditions of this Underlease (save as to the amount of rent but including the provisions of rent reviews herein contained).

1.5.6 On the assumption that the said willing tenant or tenants do not seek a rent free period nor any reduction in rent to allow them the equivalent of a rent free period and in considering any comparable rents the existence of any rent free period or any reduction in rent calculated to allow for any rent free period shall be ignored.

Fair Market Rent   In *National Westminster Bank plc v Arthur Young McClelland Moores & Co* [1985] 2 All ER 817, the phrase was defined as:

> such amount shall represent a yearly rent at which the demised premises might reasonably be expected to be let at the relevant review date in the open market by a landlord to a tenant without a PREMIUM with VACANT POSSESSION and subject to the provisions of this subunderlease other than the rent hereby reserved there being disregarded any effect on rent of any of the matters set out in paragraphs (a), (b) or (c) of Section 34 of the Landlord and Tenant Act 1954 (as amended).

Market Rental Value   In *Datastream International Ltd. v Oakeep Ltd.* [1986] 1 EGLR 98, the definition given was:

> the rent at which the said property is worth to be let in the open market as a whole at the best rent reasonably obtainable without taking any fine or premium and on the following bases:
> (i)   subject to the provisions of this lease (other than the amount of rent hereby reserved) for a term equal to the residue then unexpired of the term granted by this lease but having regard to the lessee's rights under the Landlord and Tenant Act 1954....

An alternative definition was provided in *MFI Properties Ltd. v BICC Group Pension Trust Ltd.* [1986] 1 EGLR 115, where the same was defined as:

> The rent hereinbefore first reserved shall be revised so as to equal the rent at which having regard to the terms of this sub-underlease (other than those relating to rent) [the demised premises might then reasonably be expected to be let in the open market by a willing landlord to a willing tenant for a term of twenty years with VACANT POSSESSION].

Yearly Rental Value   This is defined in *Amax International Ltd. v Custodian Holdings Ltd.* [1986] 2 EGLR 111, as 'the amount which shall in his [the expert's] opinion represent a fair yearly rent for the demised premises on the relevant date, having regard to rental values then current for property let without a PREMIUM with VACANT POSSESSION and on the provisions of this lease (other than the rent hereby reserved)'.

Net Rental Value   This was defined in *Equity & Law Life Assurance Society plc v Bodfield Ltd.* (1987) 281 EG 1448, as:

the best rent which the premises hereby demised might reasonably be expected to fetch on the open market upon the following assumptions that is to say:

(i) That they are vacant and to let as a whole without a PREMIUM or other capital payment for the residue unexpired of the term hereby granted upon the terms of this lease other than as to duration and rent.

(ii) That the premises have been kept in good repair and condition in all respects in accordance with the lessees' obligations hereunder.

As can be seen, these definitions, from various leases, are somewhat idiosyncratic but, in essence, all aim to achieve the same rental level, i.e., a rent which, were the premises to be let on the open market, vacant, by a willing landlord to a willing tenant on a fully repairing and insuring basis, could be obtained.

Fittings

These are items which may be fitted either by the landlord or the tenant: examples might be light fittings and carpets. They will normally remain in the ownership of the person who fitted them. The legal test is whether the item is fitted in such a way to the premises that it becomes attached to it or whether it is removable without damage to the building. If it is so removable then generally it will remain in the possession and ownership of the person who fitted it.

Fixtures

These are items which are attached and become part of the building or demised premises. Installation of fixtures by a tenant may result in considerable benefit to the landlord, for the reason that, once they are installed, the ownership of the fixtures passes from the tenant to the landlord. At the end of the lease, therefore, the landlord may have VACANT POSSESSION of a much enhanced building at no cost to himself. This, however, is unusual as generally fixtures will be items of plant and machinery which of their very nature have only a limited life.

FIXTURES AND EFFECT UPON VALUATION

Young v Dalgety plc (1986) 281 EG 427
By a term of the lease the landlord required the installation of light fittings and the laying of a carpet floor covering which was attached to the floor by gripper rod fittings. The question before the court, at the time of the rent review, was whether the value attached to the light fittings and the carpets should be disregarded as being tenant's fixtures.

The Court of Appeal decided that:

1. The items were required to be installed by the landlord; but unless express words were used to the contrary they remained the tenant's fixtures and were not the landlord's fixtures.

2. Therefore these items were to be disregarded in the rent review process. The carpeting was not in the floor finish but was merely attached to the floor by gripper rods for the tenant's convenience.

It is of interest to note that had these requirements been termed 'improvements', it is likely that they could have been taken into account at the review date.

Normally the clause for the implementation of the rent review requires the third party (or indeed the original parties) to make certain assumptions about the hypothetical lease in order to arrive at a rental value.

Assumptions

These are matters which the valuer, albeit acting as an arbitrator or as an independent expert, must assume to be present in a hypothetical lease granted by a willing landlord to a willing tenant for the demised premises. Normally the rent review clause will simply state that the valuer is to assume that the hypothetical lease has all the clauses that are in the actual lease other than the specific mention of the rent itself. However, where a specifically restricted USER CLAUSE is set out in the lease, it may well be beneficial to the landlord to assume that this restrictive user clause does not appear in the hypothetical lease.

If the valuer is to assume that all clauses exist in the lease other than that as to rent, then he can take into account the actual rent review clause itself, which common sense would dictate that he should. However, by reason of idiosyncratic drafting, such an assumption has seldom been clearly stated. This has brought about the result that a hypothetical lease may be assumed not to have within it a rent review clause. The implication of this in valuation terms is that the rent payable upon review takes into account the fact that there is no subsequent rent review and increases the notional rent payable by a minimum of ten per cent. This is very satisfactory to the landlord and deeply unsatisfactory to the tenant.

ASSUMPTIONS REQUIRED TO BE MADE ON VALUATION

Basingstoke and Deane Borough Council v Host Group Ltd. [1988] 1 All ER 824

There were rent review proceedings applicable to a public house or inn. The lease was for a term of ninety-nine years. The lease referred to a plot of land hereby demised. The question to be decided was what terms and conditions of the 'hypothetical lease' should the valuer take into account at the review.

The court below found that the reference to a bare site was to protect the tenant from a valuation, which took into account the cost of the works, which had been paid for by the tenant. The Court of Appeal agreed.

The other main purpose was to protect the landlord from the effects of inflation. Again the Court of Appeal agreed.

The Court of Appeal, while agreeing that many covenants were inappropriate to a bare site, directed that the following were important.

The language of the lease is of paramount importance when interpreting the review provision but the commercial purpose of the clause must always be recalled.

It would be vital in any hypothetical lease for the hypothetical tenant to know what was the permitted use and how the rent review related to that use.

Therefore the covenants which are relevant will be assumed to be contained in the hypothetical lease.

Notional lettings will be assumed to be on the same terms as the existing lease, other than the rent payable, unless express terms direct otherwise.

There was no requirement to differentiate between the rent review provisions in a ground lease as opposed to a rack rented demise.

There was no express direction in this lease to a valuer to ignore the other terms. To do so would be unfair to the tenant.

The court directed that a key reason to take account of a restricted use covenant, which benefited the tenant, was to ensure that should the landlord wish to license the occupier to use the premises for an alternative use then the review process should take account of that alternative in the assumptions.

The Court of Appeal stated that a valuer's role was confined to assessing the value of a property with stated characteristics. His role was not to decide for himself as to the characteristics unless that power were given expressly (and concisely).

These are the guidelines for the valuation process.

WHAT ASSUMPTIONS SHOULD BE IMPLIED?

General Accident Fire & Life Assurance Corporation plc v Electronic Data Processing Co plc [1987] 1 EGLR 112

There were various express assumptions in this lease. These did not include the provision to review the rent. The arbitrator determined the reviewed rent on the basis that no rent review provision should be implied.

The court agreed. It indicated that the literal approach should be followed in examination of the clauses pertaining to the review.

ASSUMPTIONS – 'ON THE SAME TERMS IN ALL OTHER RESPECTS AS THESE PRESENT'

James v British Crafts Centre [1987] 1 EGLR 139

The matter for decision was whether the USER CLAUSE which specifically named the tenant should be incorporated in the hypothetical lease.

The Court of Appeal decided that such a specific user should be incorporated. May LJ agreed that it was possible that a hypothetical open market could exist and could easily be considered on the terms of the instant lease as they stood.

ASSUMPTIONS – SHOULD A RESTRICTED USER CLAUSE BE INCLUDED?

Post Office Counters Ltd. v Harlow District Council [1991] 36 EG 151
There was one question for decision. It was to be assumed that there was a hypothetical lease. There was disagreement between the parties as to the terms to be included therein.

There was a restrictive clause in the actual lease. Having taken legal advice, the arbitrator decided that the clause in the hypothetical lease should not be on similar terms. As a result the arbitrator treated this similar clause as less onerous (which it was) and awarded a 7.5 per cent increase in the base rent. This was challenged by the tenant.

The landlord submitted that the USER CLAUSE as at present drafted would prevent the clause being incorporated into the hypothetical lease, the reason being that such an incorporation would mean that the tenant would be the only tenant in the hypothetical lease. That would be incompatible with an open market VACANT POSSESSION letting. Therefore, an interpretation of a strict nature was required in order to make commercial sense of the review clause. The court agreed with this view.

ASSUMPTIONS – EXISTING LESSEE AND HYPOTHETICAL TENANT

First Leisure Trading Ltd. v Dorita Properties Ltd. [1991] 23 EG 116
There were two separate premises. One was originally the Greyhound Hotel in Croydon. It was the subject of a lease for a hundred and thirty years from 25 March 1962 at a rent without review.

The first floor of this hotel extended over two shop units demised separately. The same parties leased to each other the two shop premises. However, although the term, i.e., a hundred and twenty-three years from 25 December 1968, was the same, there was the ability here to review the rent.

The lease for the shops was said to be on identical terms save as to the premises, the term of years, and the rent. However, before March 1989 the reversionary interest in the hotel and the two shops was split.

The shops were owned by Dorita. On 7 March 1989, Dorita gave a licence to the original tenants to assign to First Leisure. The current owners of the original hotel premises also gave licence to assign First Leisure their demise.

As a preliminary issue the parties asked the court to decide whether the actual tenant should be assumed to be a possible hypothetical tenant. The judge decided that whilst the actual tenant could not be the hypothetical lessee, the potential interest in the property by reason of the tenant's holding in the Greyhound Hotel could not be ignored. It followed therefore that the hypothetical tenants should not be restricted to those prepared to take a risk in seeking and obtaining the consent of the proprietor of the Greyhound Hotel to use the shop premises as an extension of the hotel.

It is somewhat difficult to envisage a hypothetical tenant who would not wish to bid for premises available on the same terms as the Greyhound.

ASSUMPTIONS: TERMS, RENT REVIEW PERIODS, PERSONAL COVENANT, AND A RENT-FREE PERIOD

St Martins Property Investments Ltd. v CIB Properties Ltd. and Another (11 November 1998, Court of Appeal, unreported)
The term was for thirty-five years from June 1986. The dispute concerned assumptions. The landlord, as appellant, argued that the hypothetical term should be thirty-five years commencing at the date of review. The term was to be assumed to be equal to that originally granted. The Court of Appeal directed that the only reasonable construction was that any notional term commenced in 1986 and thus upon review it was the unexpired term which was the assumption.

The next submission sought to imply rent review provisions different from reality. There was no requirement to alter the terms, as the provisions were already in place. In the actual lease the parties had agreed a covenant personal to that tenant alone, as to have a right to break the lease upon certain conditions. The landlord wished that the hypothetical lease also contained the same covenant. That was rejected. The covenant was personal and ceased upon any notional assignment.

The landlord submitted that the rental value should equate to the headline rent obtainable in conjunction with the grant of a rent-free period. That was rejected as the parties had clearly intended that the review should be carried out upon the assumption that the rent-free period should be ignored.

Fortunately express words gave the opportunity to implement the original agreement or bargain of the parties.

ASSUMPTIONS – USER CLAUSE

C&A Pensions Trustees v British Vita Investments Ltd. [1984] 172 EG 63
The rent review procedure had commenced. The landlord sought an order that the USER CLAUSE should be assumed to encompass those uses authorised by the landlord. In anticipation of the review the landlord had, without being asked by the tenant, widened the user clause.

The tenant protested that it only required the use that had been granted to it under the original lease. It was a device by the landlord to attempt to obtain a higher rent upon review.

The court, with some acerbity, ruled that the assumption should be limited to that use requested by the tenant and subsequently authorised by the landlord.

ASSUMPTION – LENGTH OF HYPOTHETICAL LEASE

Dollar Land Holdings plc v British Gas plc [1992] 1 EGLR 135
The term of the lease was thirty-five years. Rent reviews were scheduled every seven years. The clause stated: '...the full market yearly rent means the yearly rent at which the demised premises might reasonably be expected then to be

let in the open market with VACANT POSSESSION by a willing lessor by a lease in the same terms in all other respects as this present lease (other than the original rents hereby reserved but including the provisions for review of rent hereby contained)....'

The arbitrator made alternative awards for the benefit of the court. The tenant submitted that the hypothetical lease should have a term of thirty-five years from the date of review. The landlord argued that as the unexpired term was fourteen years that should be the length of the notional term.

The court recalled the *Basingstoke* (see p. 70) guideline as to what was the commercial purpose of this lease. The court directed that note should be taken of the subsisting terms of the actual lease. The court held that the phrase 'other than the original rents hereby reserved' was unnecessary. Following *Basingstoke*, the judge decided that there were explicit or implicit indications within the lease to assist the landlord. Clear weight should be given to the words 'same terms'. This meant that the notional term commenced in 1969, thus there were fourteen years to be granted for the notional lease.

This decision brought reality to a rational submission on behalf of the landlord and did not appear to change the bargain made between the parties.

ASSUMPTIONS – FLEXIBLE OR DEFINED TERM – LENGTH OF TERM

Westside Nominees Ltd. v Bolton Metropolitan Borough Council (3 February 2000, Queen's Bench Division, unreported)
The lease was for duration of 125 years from 6 November 1969. The under lease was for the same term less three days. The rent review clause was upward only every fourteen years. A formula was agreed. The premises should be valued so as to achieve the best rent to be obtained in the open market whether as a whole or in parts. The lease explicitly divided the building into two ground floor units and an upper part. The tenant argued that the divisions in the lease should be incorporated into the assumption. Alternatively the valuation should be based upon the actual division. That was not the same as described in the lease. The landlord submitted that the option identified by the use of the phrase 'in parts' indicated a flexible approach.

The court agreed that the approach was flexible.

The landlord then asked for a declaration that the hypothetical term should be some ninety-seven years at the imminent review, the equivalent of a PRESUMPTION OF REALITY. The tenant submitted that the concept of a letting of that length was unrealistic. There was little or no useful comparable EVIDENCE of any letting in excess of thirty-five years. As a result the tenant argued that the hypothetical term should be such a term as might be expected to be obtained for the particular part of the building as at the review date relying upon various assumptions.

The Court of Appeal agreed. In spite of the use of fairly explicit wording that the notional term should be as argued by the landlord the court was persuaded, to some degree, by the lack of comparables.

This is of concern. The parties agreed the terms. Is it really part of the equipment of judicial interpretation to take into account the availability of comparables in order to review the terms of the agreement? The task of valuation is for the valuer.

ASSUMPTIONS AS TO TERMS OF HYPOTHETICAL LEASE

Dennis & Robinson Ltd. v Kiossos Establishment (1987) 282 EG 857
Here there was no express mention that there should be a willing landlord and a willing tenant in the hypothetical lease where the premises were expressly deemed to be available on an open market basis with VACANT POSSESSION. The Court of Appeal was asked to decided whether such a phrase should be assumed. The decision was that:

1. There will be an implied letting of the premises.
2. There is a market for the letting.
3. There shall be an implication that the landlord is willing to let and the tenant willing to rent the demised premises.

S.I. Pension Trustees Ltd. v Ministerio de Marina de la Republica Peruana [1988] 13 EG 48
There was a rent review imminent. The question of use arose for decision. The clause stated:

> ...the expression 'the market rent' shall be deemed in this clause to mean the yearly rental value of the demised premises (having regard to rental values current at the relevant time for similar property used for any purpose within the same use class under the Town and Country Planning (Use Classes) Order 1963 as that which includes the use of the demised premises permitted hereunder let with VACANT POSSESSION without a PREMIUM and subject to provisions similar to those contained in this lease)...

The original use was as mortgage, finance, and insurance consultants. Subsequently, there was an assignment of the premises to the Government of Peru. The use was then that of a lawful diplomatic mission. At the review, the question arose as to whether the premises were to be valued as offices without a restriction as to use or not. In addition, the court was required to decide the meaning of the phrase 'market rent'.

The words 'vacant possession' were said by the tenant merely to enable the arbitrator to use vacant open market lettings to which he could then apply a discount to arrive at the rental value for the demised premises subject to their occupancy. The learned judge rejected this and determined that, as the clause did not require a discount to be applied, the rent should be on an open market value with an assumption of vacant possession.

As to the effect of the restriction upon use and its effect upon the rent review proceedings, the learned judge decided that the value must be on the basis of the terms just stated but limited to a use equivalent to that of the original tenant, i.e. mortgage, finance, and insurance consultants.

Wallace v McMullen & Sons Ltd. [1988] 28 EG 81

In the lease of premises which constituted a golf club, the rent review clause required the value to be that as to capital and not rental. The rent was then to be determined by a percentage emerging from the difference between the original freehold value and the actual value at review. The value was to be assumed to be subject to rights and obligations together with planning consent for a golf club. The question was whether the land was to be valued subject to all the obligations arising as between landlord and tenant or merely as between adjacent freeholders with VACANT POSSESSION.

The first decision, made by the learned Vice-Chancellor, was that the land was to be valued as agricultural land, with the benefit of planning permission for a golf course, as this was the situation when the lease was granted. Secondly, and interestingly, it was decided that the land was to be valued with vacant possession in spite of the fact that there was no clear statement to that effect in the lease. This is surprising as, even in 1971, when the lease was drawn up, the necessity to visualise a property as vacant and to let was treated as the norm.

ASSUMPTIONS AND EFFECTS UNDER LETTING OR VACANT POSSESSION

Laura Investment Company Ltd. v Havering London Borough Council (no. 2) [1993] 08 EG 120

There was formerly a bare site. It was to be let to a tenant. This tenant would construct buildings and underlet the same for industrial and warehousing purposes.

The review clause had an assumption that all covenants and conditions for lease had been performed. The building was yet to be erected at the time of the granting of the lease and was to be included in the hypothetical demise. The question for the court was: were the buildings to be assumed to be let or vacant? The case in favour of the VACANT POSSESSION assumption is that such an assumption would mitigate against the tenant who, by taking a PREMIUM and letting the actual premises at an artificially reduced rental, thereby reduced the hypothetical rental value, this by reason of the difficulty of deciding the value of a premium.

Here underlettings were required by the lease. Consent to underlet could not be unreasonably withheld.

As it would be uncommercial for the tenant to agree to a condition of the lease which required the tenant to pay rent for premises erected at its own cost, the court decided that the buildings should be deemed to be underlet, as they in fact were. This result seems fair as the assumption presumes that all the obligations are met. The main obligation here was to build and then underlet.

ASSUMPTION – VACANT BUT FIT FOR IMMEDIATE OCCUPATION AND USE

Pontsarn Investments Ltd. v Kansallis-Osake-Pankki [1992] 22 EG 103
The question before the expert was the definition and implications resulting from the meaning of the phrase 'fit for immediate occupation and use'.

The expert heard the submissions of the parties. He concluded that the phrase intended that in arriving at the rental valuation no allowance should be made for the period normally allocated for fitting out by the tenant. In other words, the premises were deemed to be ready for immediate occupation on the assumption that all fitting out had already taken place.

The court decided, supporting the clear decision in *Jones v Sherwood* (see p. 90), that it could not permit a challenge to the expert's DETERMINATION. However, it proceeded to decide the meaning of the phrase. This decision was unnecessary to judgment and should not, in my view, be treated as more than a persuasive indication of judicial thinking. This suggests that the building is fit for immediate use even where the fitting-out work has yet to take place. If a residential unit is considered as an example, then fit for immediate use suggests that a family could move in. However, this was not the view of the learned judge.

ASSUMPTIONS – FIT FOR IMMEDIATE OCCUPATION AND USE OR RENT FREE AND FITTING OUT BY TENANT?

London & Leeds Estates Ltd. v Paribas Ltd. [1993] 1 EGLR 121
The lease was for a twenty-five-year term with rent reviews every five years. The disregards were the term and the amount of rent but excluded the provisions for the rent review. The assumption was phrased 'That the demised premises are fit for immediate occupation and use...'. The judge noted that the issue arose out of the frequent situation where landlords, very wisely, offered premises to let in a 'shell' form, allowing the tenants to fit out as they wished. The tenants were offered a variety of financial inducements to persuade them that occupancy was an attractive proposition.

Leases were prepared which attempted to remove any risk of a double benefit arising for either party. Both parties agreed to reject an approach which treated the premises at the review date as being in a 'shell or developer's finish'.

The tenant submitted that the works of fitting out would have been completed at his own expense prior to the letting. The court rejected that view.

The landlord agreed that the fitting out works had been completed but that there was no assumption as to who will have paid for the works. The court accepted that view. The landlord's view was the only one that did not require the need to imply an additional term. Further the parties had intended that the bargain would produce a situation whereby the cost of the fitting out equated to the amount of rent forgone by the landlord.

A sensible conclusion but one which ignores the situation where inducements are given to occupy premises in a falling market or where very favourable terms are given in order to secure an anchor tenant. Whenever this arises the tenant is well advised to add a rider to the assumption provision setting out the exact situation applicable to the inducement.

ASSUMPTIONS – MEANING OF 'PROPORTION'

Stylo Barrett Properties Ltd. v Legal and General Assurance Society Ltd.
[1989] 31 EG 56
There was an agreement to grant a lease for forty-two years. Thereafter there were two leases, the second commencing at the expiry of the first but both leases made on the same date. The second lease was subject to a rent review.

The assumption was that the new rent was that proportion of the fair rack rental value (which was the best rent obtainable from a willing tenant or tenants whether as a whole or in parts...), which the passing rent bore to the initial rent immediately after the date of the review.

The original rent payable was 109.13 per cent of the initial rent. The rent payable was £49,000 compared with an initial rent of £44,900. The tenants argued that the new rent should remain at £49,000. If not, then the tenants could be required to pay a sum greater than the gross rental value if the formula were implemented as semantically required. The tenants sought a declaration that 'proportion' would exclude a figure equal or greater than the whole.

The court decided that the word should be interpreted upon its ordinary English meanings. While proportion did have the meaning that was contended by the tenants, it also had the alternative meaning. There was no reason why the parties should not have agreed a formula which resulted in the rent payable being greater than the rental value. That was the order of the court.

As has been stated before, the parties made that bargain, and perhaps the word 'proportion' should have been defined in the lease. However the difference between the rent passing and the initial rent made the landlord's argument logical and fair.

Disregards

Disregards are those matters which the valuer upon rent review must not take into account when determining the new revised rental value. As with assumptions, they will normally be clearly set out in the lease and will include the fact of the tenant's existence in the premises; any GOODWILL attaching to the presence of the tenant and his business; and any improvements which the tenant has carried out for which he has sought approval and received the landlord's consent.

If the position of the tenant is going to be adequately protected at the review it is important to arrive at a value for the improvements, as well as to determine that the correct procedure is followed at every instance. The fact of the tenant's presence in the premises and also of any goodwill attaching to his presence are usually easily valued. The valuer simply treats the premises as vacant and to let. This does not therefore put the valuer to great mathematical stress and, indeed, makes use of EVIDENCE of comparables relatively easy.

There are, however, problems when the lease either does not state whether there are any disregards at all or limits the number of disregards or merely makes reference to those disregards as set out in the Landlord and Tenant Act 1954, s. 34 (1), the essence of which is as follows:

1. Any effect on rent attributable to occupation by the tenant (or a predecessor in title) of the holding. (The tenant is not, therefore, protected against the open market rent by the fact that he is a sitting-tenant.)
2. Any goodwill attaching to the premises by reason of the business carried on by the tenant or any predecessor in title of his in the same business.
3. Any increase in value attributable to certain improvements carried out by the person who was current tenant other than in pursuance of an obligation to his immediate landlord. The following conditions apply to this, by s. 34(2):
   (a)   that the improvement was completed either during the current tenancy or not more than twenty-one years before the application for a new tenancy;
   (b)   that the holding or the part improved was at all times since completion of the improvement subject to tenancies to which Part II applies;
   (c)   that on the termination of any tenancy the tenant did not quit.
4. In the case of licensed premises, the increase in value attributable to the licence, if its benefit is to be regarded as the tenant's.

DISREGARDS AND ISSUE ESTOPPEL

Arnold v National Westminster Bank plc [1988] 45 EG 106
There was a clause in the lease that required the reviewed rent to be determined 'subject to the provisions of this under-lease other than the rent hereby reserved'. There was an arbitration as to the value of the rent on review. The arbitrator decided that the lease should be treated as if there were a five-year review pattern. He also stated that if he were wrong and there were no rent review in the hypothetical lease, the rent should be twenty per cent higher.

The landlords appealed against that decision. Walton J, the judge at first instance, agreed that the words should be defined as excluding any ability to review the rent in a hypothetical lease. He refused LEAVE to appeal by the tenants to the Court of Appeal.

The date of the new rent review approached. The tenants sought to have a decision upon a point already decided at the previous review. There was no question that in the circumstances the principle of issue ESTOPPEL applied: the matter had already been litigated. The tenants claimed, however, that because the Court of Appeal had come to a different view from the trial judge in similar cases, this circumstance should be one of those where issue estoppel would not prevent relitigation.

The Vice-Chancellor reminded the parties that judicial opinion was divided. He supported the view that rent review should be considered not in isolation, but as a clause which has the intention of attempting to give to the landlord upon review the current market rent. It was stated that the trial judge held the opposite view.

It was decided that there had been a change in the law. The principle of issue estoppel is subject to exceptions. It is not a rigid rule. 'A change in the law can constitute further material which is relevant to the correctness or incorrectness of an earlier decision.' The key element was whether this change in the law and this earlier decision brought injustice to one of the parties. The following criteria are set out from the judgment.

1. There is a continuing contractual relationship of landlord and tenant under which (if there is an issue estoppel) the decision of Walton J would regulate four further rent reviews and thereby affect the rent payable until the end of the term.
2. Because of the peculiarities of the procedure applicable to appeals from arbitrators, unlike the ordinary case of a prior decision by a judge, the decision of Walton J was not subject to appeal. Therefore a matter of very great financial importance would, if an issue estoppel applied, be decided on a POINT OF LAW which the lessees had never had the opportunity to test in the higher courts.
3. The decision whether or not to permit an appeal was the decision of Walton J himself and there was no right of appeal against his refusal to certify the matter fit for appeal. The lessees took every possible step to test the earlier decision in the higher courts but without success.
4. Subsequent decisions, in particular that of the Court of Appeal in the *Equity & Law Life* case (see p. 82), made it, at the lowest, strongly arguable that the decision of Walton J was wrong.

Leave to have the matter retried was granted to the tenants.

DISREGARDS AND MEANING OF 'OTHER THAN AS TO THE YEARLY RENT'

British Gas Corporation v Universities Superannuation Scheme Ltd. [1986] 1 EGLR 120
Amongst the questions for decision was whether the phrase 'other than as to the yearly rent' meant that all the provisions as to rent should be excluded from the hypothetical lease. The learned Vice-Chancellor decided that only the rent payable should be excluded. Guidelines were given for the construction of clauses:

(a)   words in a rent exclusion provision which require all provisions as to rent to be disregarded produce a result so manifestly contrary to commercial common sense that they cannot be given literal effect;

(b) other clear words which require the rent review provision (as opposed to all provisions as to rent) to be disregarded must be given effect to, however wayward the result;

(c) subject to (b), in the absence of special circumstances it is open to give effect to the underlying commercial purpose of a rent review clause and to construe the words so as to give effect to that purpose by requiring future rent reviews to be taken into account in fixing the open market rental under the hypothetical letting.

The question which arises is whether a valuation would be possible if the current rent were not excluded. What the valuer is usually being asked to value is a premises vacant and to let. Therefore, the fact that an historic passing rent is currently payable is irrelevant to current value.

DISREGARDS – THE EFFECT UPON RENT OF A PARTIAL CLAUSE ('SAVE AS TO RENT')

British Home Stores plc v Ranbrook Properties Ltd. [1988] 1 EGLR 121
Here was a dispute about the meaning of the phrase 'save as to rent' given in guidance to the valuer of a reviewed rent as to the assumptions and disregards which should be taken into account.

The court decided that the guidelines, described in *British Gas Corporation v Universities Superannuation Scheme Ltd.* (see case above), must be relevant. Therefore, it was required to examine the commercial purpose of the process and of the contract. The tenant was to have security of tenure subject to adherence to the covenants of the lease, and the landlord in return was to have the right to have the rental return regularly increased by means of the review procedure. In determination of the meaning of words or phrases the wide view should be taken. The test was decided to be as to whether the phrase could have only one meaning. In this instance the phrase was ambiguous. Therefore, the guide must be the commercial effect, which was to achieve a review taking into account the fact of review. On that basis the phrase excludes only the amount of rent payable.

A DISREGARD – THE MEANING OF 'THE RENT RESERVED'

Telegraph Properties (Securities) Ltd. v Courtaulds Ltd. (1980) 257 EG 1153
In the original lease the clauses and disregards were carefully set out. In the subsequent lease there were merely references to the original lease but with certain sub-clauses specifically excluded.

The phrase 'the rent reserved' was expressly to be disregarded from a lease to adjacent premises which were taken from the same landlord after a surrender of part of the premises originally demised. That phrase, by implication, excluded all matters pertinent to both the rent and the rent review clause. In the original lease there was a rent review clause.

The question for the court was: did the phrase 'rent reserved' exclude the rent passing alone and include everything else? It was held that it did.

DISREGARD – EXCEPT AS REGARDS RENT

Electricity Supply Nominees v FM Insurance Company Ltd. [1986] 1 EGLR 143

The rent review machinery directed that all the terms of the lease were to be assumed except as regards rent. The tenants argued that the term intended to exclude only the amount of passing rent at the time of the review. The landlord argued for a wider meaning.

The court found that the purpose of the review clause was to ensure that the rental value did not fall in real terms during the period of the tenancy. The phrase in question could exclude such covenants as the requirement to pay rent; the review provisions themselves; and the provision that allowed the landlord to re-enter the premises in the event of non-payment of rent.

That approach would give a meaning that was inconsistent with the commercial purpose of rent reviews.

The court decided that the term identified only the passing rent and nothing more.

It might not be the 'commercial purpose' but that might have been the intent at time of signing the agreement. Whether courts would adopt a stricter interpretation now is arguable.

EFFECT OF PHRASE 'OTHER THAN RENT HEREBY RESERVED'

Amax International Ltd. v Custodian Holdings Ltd. [1986] 2 EGLR 111
The court decided that:

1. The phrase 'other than rent hereby reserved' does not mean:
   (a)   All matters set out within the rent review clause.
   (b)   Nor is the arrangement of the phrase such as to cast doubt upon the inference that the rent review clause was intended to be incorporated in any hypothetical lease.
2. The words are limited in meaning to 'the amount of rent payable'.
3. As the phrase does not require the removal of any part of the lease provisions, it is superfluous and adds nothing.

If the judicial view is, as it appears to be, that the phrase adds nothing to the interpretation or meaning of the clause, then one might ask why the phrase was inserted.

MEANING OF 'OTHER THAN AS TO DURATION AND RENT'

Equity & Law Life Assurance Society plc v Bodfield Ltd. (1987) 281 EG 1448
The Court of Appeal stated that the phrase 'other than as to duration of rent':

1. did not exclude a covenant to pay rent;
2. did not exclude a covenant to re-enter for non-payment of rent;
3. did exclude provision for the machinery of review because provisions as to discounts in the first period of the tenancy could not be introduced into the hypothetical letting.
4. Dillon LJ, with whom Fox and Russell LJJ unreservedly agreed, welcomed and approved the guidelines indicated by the Vice-Chancellor in *British Gas Corporation v Universities Superannuation Scheme Ltd.* (see p. 80), but commented that these were only guidelines. The function of the court was to construe, in each particular case, the rent review clause which was in use.

DISREGARD – FAIR RACK RENTAL MARKET VALUE & ADDITIONAL RENT?

Lister Locks Ltd. v TEI Pension Trust Ltd. [1981] 264 EG 827

> Upon review the revised rent shall be the greater of either the previously payable rent or the fair rack rental market value ... together with the extra reviewed rent. The fair rack rental value ... shall be the best rent at which the premises might reasonably be expected to let for a term of years equivalent to the then unexpired residue ... in the open market by a willing landlord to a willing tenant and subject to similar covenants and conditions (other than the amount of rent) ...

The court was required to determine whether one of the covenants, which was not excluded, was the requirement to pay the 'extra reviewed rent'. In other words, was the expert to disregard the fair rack rental value only or the accumulation of both sums? The extra rent was determined by means of a concise formula.

The court took account of the circumstances, which led to the agreement as to the terms of the lease. The correspondence was explicit as to the awareness of the tenant that the rent would be the combined figure. The court ordered that the disregard should be the combined figure.

DISREGARDS – FACT OF OCCUPATION – MEANING OF 'EXCLUSIVE RENT', 'MARKET RACK-RENTAL VALUE'

Hill Samuel Life Assurance Ltd. v The Mayor, Aldermen and Burgesses of the Borough of Preston [1990] 36 EG 111
This was a ninety-nine-year lease. The ground rent was a peppercorn. Later the rent was adjusted to £28,000 per annum. There were provisions for rent reviews every fourteen years. The parties agreed that the lessee held the lease as an investor not as an occupier. The court ruled that that fact was important in considering the assumptions. The intent of the parties was that the reviewed ground rent should bear the same proportion to the rack rental value as was fixed by the initial rent. The initial rack rental figure was agreed at £172,000

but the ground rent was £28,000 per annum. What were to be the assumptions and disregards?

The lessees erected a multi-purpose building. It was always clear that the building should be let in parts. The disregards included the facts of the lessees' occupation, the value of any rental placed upon the premises by reason of any business carried on there, the effect of improvements carried out after the date of the agreement except under an obligation of the lease and the value attributed to fixtures and fittings.

The definition of 'exclusive rent' was the sum of the rents obtainable on a letting of the individual parts to the intended sub-lessees. The definition of 'market rack-rental value' was required in the light of above finding. The court held that the premises must be valued as if vacant and relied upon the disregard as to the non-existence of the sub-lessees for support. The valuation was to be determined on the basis that the property was held as an investment and not upon the basis of a willing tenant considering vacant premises.

The valuer would then determine the occupation rent for either the individual units or the premises taken as a whole.

A practical solution to a very difficult valuation exercise.

MEANING OF 'SAVE FOR THIS PROVISO'

Safeway Food Stores Ltd. v Banderway Ltd. (1983) 267 EG 850
There was a rent review clause requiring the valuer to consider hypothetical letting on identical terms and conditions 'save for this proviso'.

The court decided that this phrase referred to the entire rent review clause, and therefore the hypothetical lease should be deemed not to have such a clause.

DISREGARDS AND EFFECTIVE RENT-FREE INDUCEMENT

City Offices plc v Bryanston Insurance Co Ltd. and *City Offices plc v Allianz Cornhill International Insurance Co Ltd.* [1993] 11 EG 129
The relevant element in the disregards was phrased as follows: 'any notional rent-free rent concession or fitting out period for which allowance would or might be given to the tenant if the demised premises were let in the open market with VACANT POSSESSION'.

The arbitrator, who took legal advice, decided that in March 1992 a rent-free period of twelve months would have been allowed. He made a deduction of three months for that period for fitting out. He allowed a discount of 15 per cent of the rental value for the rent-free period. The landlord objected and appealed to the court.

He claimed that the phrase: '... and for this purpose the current market rental value shall mean the best yearly rent which would reasonably be expected to be payable ... after the expiry of any rent free concession or fitting out period ...' meant that the value would be the rent payable for the residue of the unexpired term before the next review or end of lease. Thus, the disregard as to the notional rent-free concession intended the assumed or hypothetical term to commence as if no rent-free period had ever been granted.

The learned judge disagreed. This would be excessively advantageous to the landlord. It took no account of market conditions, changes in property values, or the value of money. The court decided that the lower figure which benefited the tenant was correct.

Here the landlord had attempted to ensure that, when markets improved, the value of his investment would be preserved. While it is possible to regret the lack of clear language to obtain this end, it is felt that the learned judge was incorrect. If the property market had risen the landlord would have benefited by means of any increase in demand. In a falling market he is penalised. It is suggested that an appeal would not have been unwise in this situation.

DISREGARD – INDUCEMENT PERIOD

Scottish Amicable Life Assurance Society v Middleton and Others [1994] 1 EGLR 150

An arbitrator had been appointed. The parties agreed to seek the answers to various questions.

The review clause directed that the rent was to be the best yearly rent following the expiry of any rent-free period or period of concessionary rents. The question was whether the rent-free period was limited to such periods as a willing landlord would give to reflect the fact that the lessee would not be able to make full use of the premises until the completion of his own fitting out works. The court noted that the exclusion of rent-free periods gave a bonus to landlords upon review. That would increase the rent artificially.

Also the DETERMINATION of the open market rent at the rate payable following the expiry of the inducement period is not part of a process, which reflected neither the change in the value of money nor real increases in the value of property.

The court directed that the market rent was to be at the rate payable following the expiry of the inducement period. The valuer was required to make any adjustment necessary to eliminate the effect on rent attributable to the tenant's need to give an inducement to any subtenant.

This is still a very contentious issue not only in terms of rent review but also in terms of valuation of the reversionary interest.

DISREGARD – LANDLOCKED LAND

J. Murphy & Son Ltd. v Railtrack plc [2002] 32 EG 99

Here the property was landlocked. The lessee sought a renewed tenancy under the Landlord and Tenant Act 1954 Part II. The landlord submitted that the reality of the land being landlocked should be disregarded. The court below disagreed and accepted a discount of 40 per cent. The expert for the landlord offered one third discount and that of the tenant suggested two thirds.

The Court of Appeal decided that the criteria for ignoring the reality stated in *British Airways* (see p. 1) did not apply. The statutory provisions did not permit the court to take the background information known to the parties at the time of the agreement into account when making the ruling.

The concept of permitting a discount was upheld.

Any tenant taking occupation of a piece of landlocked land must insist that the fact is noted in the preamble. Further if the parties can agree as to the appropriate discount at the time of signing the agreement that would be the best practice.

Improvements

In the eyes of the general public, improvements are such building works as will result in a building which is of greater value, or granting greater amenity, to the inhabitant. It is also the general view that both landlord and tenant benefit by carrying out improvements. This is not always the case: see, for instance, *Pleasurama Properties* (see p. 87), where the landlord did not consider that, of its own nature, a dolphinarium improved the premises and drafted the lease in such a way as to require the tenant to reinstate the building in its original form at the end of the lease.

When improvements are carried out by the tenant it is essential that the procedure set out in the lease or by Act of Parliament, relevant to improvements, is strictly adhered to, the reason being that in most leases, and under the Landlord and Tenant Act 1954, tenants' improvements which are authorised in the proper form by the landlord are not taken into account as to their value, in certain circumstances, when the rent is reviewed to current market value. On the other hand, improvements which are carried out at the request or instruction of the landlord are not normally disregarded when taking into account the value of the premises.

It may be that an office premises before improvements is only worth £10,000 per annum. However, the same premises with the improvements carried out could be worth £15,000, i.e. a difference of £5,000 per annum. If the consent of the landlord is not sought before the improvements are carried out and its authorisation properly received, and, in certain instances, an application to the court made, then it could be that the value of those improvements will not be removed from the notional value of the premises. Thus the tenant will be penalised for improving his own premises. This state of affairs should never occur if the tenant is properly advised, or if he carefully follows the terms of his lease. However, it is surprising how many tenants blatantly ignore the procedure and then wonder why their rent reflects the improvements – in their view quite unjustly.

It is useful if the approved plans for improvement, together with the landlord's consent, are attached to the lease. It is also recommended that both the capital value of the improvements and the increased rental value that these improvements will have upon the premises is agreed between the parties at the time when the improvements are carried out. The agreements should also be annexed to the lease. In any claim for compensation at the end of the lease the fact that this has been done may save considerable cost to the tenant in attempting to value the works some years after they were carried out.

IMPROVEMENTS CARRIED OUT UNDER AN EARLIER LEASE

Brett v Brett Essex Golf Club Ltd. (1986) 278 EG 1476

There was disagreement as to whether improvements carried out during an earlier lease should be disregarded. The question for decision was whether a reference to s. 34 of the Landlord and Tenant Act 1954 Part II, set out in a lease granted after 1969, meant the unamended 1954 Act, or whether it implied the incorporation of the amendments of the 1969 Law of Property Act. In the latter Act, improvements are to be disregarded when carried out not more than twenty-one years before the application for a new tenancy.

The decision of the Court of Appeal was that the improvements were to be taken into account. The reasons were:

1. Improvements can only be improvements to demised premises, but in this case all 'improvements' were already part of the demised premises at the time of the grant of the new lease. Therefore improvements could not be separated from the premises.
2. There was no common law presumption that an amending Act shall be implied where a lease was agreed at a date after which the amending Act had come into force, but which Act was not referred to in the lease. To be certain, a precise examination of the reference is required. Here there was no reference to s. 34 (2) which would be necessary in order to include the provision as to twenty-one years.
3. Furthermore, where there were other amendments to Acts these were specifically referred to, a further implication that where no reference was made none should be implied.

VALUATION – IMPROVEMENTS AND OBLIGATION TO REINSTATE

Pleasurama Properties Ltd. v Leisure Investments (West End) Ltd. (1986) 278 EG 732

The tenants sought permission from the landlord to create a dolphinarium. This permission was granted. Improvements in the lease were to be disregarded in assessing the reviewed rent. This was an improvement.

The tenants made an appeal against a decision that the obligation to reinstate the premises at the end of the term should not be treated as an onerous obligation to be taken into account when determining the reviewed rent.

The Court of Appeal dismissed this appeal stating that if the benefit of the temporary improvements were to be disregarded it would be wholly logical that the burden of reinstatement should not be taken into account.

The hearing, in whatever form, has ended. One party wishes to submit further or fresh evidence which has come to his notice.

Fresh Evidence Available after the Award or Determination

It is rare, but not unknown, that one party may unearth fresh EVIDENCE apparently vital to their case after the end of the relevant hearing, but before the publication of the award. In that event the party who has discovered the evidence should apply to the arbitrator for an opportunity to put that evidence before the arbitrator (the same would apply to an independent expert). The arbitrator (or expert) should then notify the other side as to the request. Where the other side does not concur with the reception of any fresh evidence then the arbitrator (or expert) should call an early meeting of the parties' representatives to hear argument from both sides upon the matter. The arbitrator should only consider allowing such evidence to be put forward where:

1. It is vital to the party submitting it;
2. It was not available at the time of the hearing in spite of all efforts that reasonable professional men or women would take to ascertain the aforementioned evidence.

It may well be that the arbitrator, or expert, should the above criteria be fulfilled, will grant permission that such evidence may be put forward but may indicate that the costs attached to such a hearing should be borne by the applicant whatever the outcome.

The application to submit fresh evidence has been permitted or refused. The new rental level has been determined – the question sometimes arises as to when and from what date interest is payable.

INTEREST ON SHORTFALL OF REVIEWED RENT

Shield Properties & Investments Ltd. v Anglo-Overseas Transport Co Ltd. (no. 2) (1986) 279 EG 1088
In this case there was a claim for outstanding interest. There was no specific clause in the rent review machinery as to payment of interest, but there was a general clause within the lease indicating that interest would be payable by a tenant in default.

The court held that interest did not run until the rent newly determined was payable. Thus any shortfall due commenced from the date of review but was not payable until the quarter day after DETERMINATION. Interest was not due unless that shortfall were not paid on the due quarter day.

The judge also indicated that, where there was uncertainty, a solution to the problem might be that, when the tenant issued a NOTICE OF MOTION for the guidance of the court in the matter, the landlord could seek an order that put the tenant on terms as to payment of interest if the notice of motion went ahead.

An alternative solution is for the draftsman of the lease to consider this obvious point and ensure that the compilation of the rent review clause concisely includes an element as to when interest is payable on any shortfall.

The award, or determination, has been published. The losing party may want to attack all or part of it. The procedure for attack varies and depends upon the exact status of the third party deciding the new value.

Appeal from Determination by Independent Expert

An Award without any Reasons (A Non-speaking Award)

There is hardly any opportunity to appeal against this DETERMINATION unless it can be independently shown that the expert was biased or unjust or in some way acted to the prejudice of the parties.

Campbell v Edwards [1976] 1 All ER 785
Here there was a lease of a flat. If the lessee wished to assign there was a requirement to offer the lease back to the lessor. The value of the unexpired term, in that event, was to be determined by a chartered surveyor appointed by the President of the RICS.

The lessee wished to assign. The parties eventually agreed that Chestertons should provide the valuation. Chestertons valued the unexpired term at £10,000. The lessee gave up possession. The lessor disputed the price. Additional valuations were obtained at £3,500 and £1,250. The lessor refused to pay the previously determined £10,000 on the ground of negligence. The lessee counter-claimed for the £10,000.

The Court of Appeal decided that the valuation was binding. This was an award without reasons and, as such, was unable to be attacked.

1. There is no appeal from a NON-SPEAKING AWARD unless fraud or collusion are alleged.
2. In such an attempted appeal, joinder of the valuer is not permitted, for if joinder were to be permitted then DISCOVERY of the valuer's notes and other matters would be permitted. An investigation of these notes might then allow an appeal to lie against the valuer on the ground of negligence, which cannot be what the parties intended.
3. There is extreme difficulty in SETTING ASIDE a valuation when the parties have irreconcilably altered their positions. In this instance, one of the parties had gone out of possession of the property in question.

Appeal from an Award which has Reasons attached to it (A Speaking Award)

There appears to be no doubt that where a SPEAKING AWARD by an expert is shown to have been made on 'a fundamentally erroneous basis', then an appeal to the court can be made against such an award.

From the judgments of the Court of Appeal in *Heyes v The Earl of Derby* (1984) 272 EG 935, it seems that the only way that an appeal may arise from such an award is when the award is acted upon by the other party. Thus, in the *Heyes* case the tenant sought enforcement of the judgment for damages awarded by the valuer. The landlord sought to have the judgment put aside on the grounds that the valuation was fundamentally erroneous. Whether the landlord could in fact have appealed the valuation in any event, i.e. on the grounds of the independent expert's negligence, is another question.

APPEAL FROM SPEAKING AWARD OF INDEPENDENT EXPERT

A. Hudson Pty Ltd. v Legal & General Life of Australia Ltd. (1986) 280 EG 1434

This was an appeal against the valuers' inclusion of certain measured areas within their valuation.

The appeal was rejected as there was no discernible mistake in the valuation. The Privy Council decided that they were not concerned to consider the kinds of mistake which might justify interference by the court. Furthermore, they felt that 'attacks on valuations by experts, where those attacks are based upon textual criticisms more appropriate to the measured analysis of fiscal legislation', should be discouraged.

Campbell v Edwards [1976] 1 All ER 785

An *obiter dictum* from the above judgment was that if a valuer gives a speaking valuation and the calculations could be shown to be wrong, then that valuation might be able to be upset.

CHALLENGE TO A SPEAKING OR A NON-SPEAKING AWARD

Jones v Sherwood [1992] 1 WLR 277

The Court of Appeal was asked to decided when a DETERMINATION was final and binding.

The parties had a contract for the sale of the share capital of a public liability company. Within the terms of the contract was a clause that they could appoint an accountant as an independent expert to determine the matter in dispute between them.

It was necessary under the contract to value the shares. If the parties could not agree, then the determination clause would be implemented. The parties could not agree. The determination was made. One party wished to challenge the determination. The court decided that the expert had followed the instructions of the parties.

The determination of the expert gave no reason for its decision. It was not required to do so. The challenge was made on the grounds that the determination was the result of a mistake.

The Court of Appeal concluded that there was little difference, if any, between a speaking and a non-speaking determination. In order to decide the validity of a determination various legal investigations must be made. The first is to examine the agreement between the parties. What did the parties require of the expert? Then it is necessary to decide whether there is any evidence of a mistake. Where a mistake is a departure from instructions then a challenge would be possible. However, where there is no evidence of a departure from the instructions then no mistake will exist and no challenge is possible. Therefore, in this case, there was no successful challenge.

This is an important judgment. It disapproves of the judgment in *Burgess v Purchase and Sons* [1983] Ch 216, and it reaffirms that it may be possible to join a valuer as a defendant in order to obtain DISCOVERY of his reasons. Further, the judgment also comments that it may be possible to obtain an order against the valuer to answer a series of interrogatories (questions) where his reasons are insufficient but they have indeed been given.

APPEALS AND EXPERT DETERMINATIONS

Nikko Hotels (UK) Ltd. v MEPC plc [1991] 28 EG 86
An expert had made a DETERMINATION. The tenants were upset at the decision and challenged it.

The court decided that the rules applicable were those set down in *Jones v Sherwood* (see case above). The criterion was to adjudge whether the expert had performed the task he had been given. The expert had been asked to value and assess the average charge of each bedroom in a hotel. This would include rooms sold on a discount to agents and airlines. The expert decided that the average charge for rooms was based on the published rate and occupancy numbers.

The learned judge decided that the instructions to the expert were simple: to investigate room charges and compute the result based upon the occupancy details. Anything more than that would be outside the scope of the inquiry. Therefore the expert was not mistaken in his approach and had not gone outside his instructions. The determination was not able to be challenged.

One point to note here is that the expert decided on his own volition to employ counsel as a leading expert. He did inform the parties but did not apparently, as is normal, seek their permission. It could be, of course, that he had reserved this power to himself in the ORDER FOR DIRECTIONS.

EXPERT DETERMINATION RENDERED INVALID – MUST OBEY INSTRUCTIONS

Veba Oil Supply & Trading GmbH v Petrograd Inc [2002] 1 Ll R 295
Experts were appointed to make a DETERMINATION. The challenge to the determination was that there was a material departure from the instructions given so as to invalidate the finding. The court found that there had been a material departure.

The court found that a valuation that was erroneous in principle equated to a departure from instructions. With great respect, where there are esteemed but opposed views as to valuation concepts, then that *obiter* finding may require reappraisal.

The court defined various elements. A departure is where the expert does not carry out his instructions. A mistake made while carrying out instructions properly will not invalidate the determination.

The key is that once the court has established a material departure then there is no need to be concerned with effect. As to manifest error, the definition is oversight and blunder, so obvious and obviously capable of affecting the determination as to admit no difference of opinion.

Dyson LJ added that a term should be implied into any contract as to the effect upon a determination of any material departure.

What has yet to be clearly examined and which occurs regularly is a material departure from the instructions given by an expert or an arbitrator to the parties and the tribunal itself. It is suggested that they should not be varied without written consent from all concerned.

THE FUNCTION OF AN EXPERT TO MAKE DECISIONS

Norwich Union Life Insurance Society v P&O Property Holdings Ltd. and Others [1993] 13 EG 108

An expert was appointed to decide matters of construction of a funding agreement. A dispute arose as to the completion date of a development. However, one party alleged that, when the agreement was made, it was not envisaged that the expert would decide matters of legal interpretation.

The court decided that the agreement was straightforward and all-encompassing. The parties wished the expert to decide the completion date. In order to come to this decision the expert would have to make findings of construction or interpretation. Non-legally qualified experts such as surveyors are frequently asked to decide mixed questions of law and fact. Therefore the judge found that the expert was permitted to make an all-encompassing DETERMINATION. The Court of Appeal agreed and confirmed, as had *Jones v Sherwood* (see p. 90), though it was not cited, that the courts could make a finding on a particular point for the assistance of an expert prior to his making his determination but not afterwards.

POSSIBLE IMMUNITY OF INDEPENDENT EXPERT FROM ACTION FOR NEGLIGENCE

Palacath Ltd. v Flanagan (1985) 274 EG 143

The lease in question required the DETERMINATION of the reviewed rent to be carried out by an expert, not an arbitrator. The question to the court was whether the independent expert was immune from an action for negligence. The court held that he was not immune from such an action:

1. The onus is on the expert to establish that he is in the same legal position as to immunity as an arbitrator.

2. It was considered that the expert was generally following the criteria set out in *Arenson* (see p. 94), but a major area of omission was the requirement to act judicially.

He was not obliged to make any finding or findings accepting or rejecting the opposing contentions. Nor, indeed, as I see it was he obliged to accept as valid and binding upon him matters upon which the parties were agreed. He was not appointed to adjudicate on the cases put forward on behalf of the landlord and the tenant. He was appointed to give his own independent judgment as an expert, after reading the representations and valuations of the parties (if any) and giving them such weight as he thought proper (if any). That being so, there can be no basis for conferring immunity upon the defendant ...

The procedure for appeal is different where the determination is made by an arbitrator. It is useful to define the role of an arbitrator and when an action for negligence might arise.

Arbitrators and Liability for Negligence

There continues to be much debate as to precisely what is an arbitrator and consequently what is the liability for negligence appertaining to an arbitration.

As far as rent review procedure is concerned, an arbitrator or independent expert will not be appointed generally until there has been a service of notice as to the revised rent by the landlord and service of a counter-notice by the tenant objecting to the proposed rent and making an application to arbitration. That is without doubt a mutual failure to agree. It must surely be deemed to be a dispute under the Arbitration Act. And where there is such a dispute then the arbitrator, appointed as a result of such a dispute, should be unable to be sued for negligence during the existence of the arbitration.

However, this theory has not been precisely tested in the courts and there is uncertainty – mainly for the reason that the arbitrator is engaged by the parties whereas the judge is appointed and paid by the state. Furthermore, the supposed immunity from negligence for both arbitrator and independent expert has been cast in doubt by the judgments in *Sutcliffe* and *Arenson* discussed briefly below.

There is a possible solution, although initially it is likely there would be considerable resistance. The idea was proposed by Lord Kilbrandon, in the latter case, that it could be a term of the reference that immunity from suit for negligence would be given to the arbitrator. Is there any reason why reference to an independent expert should not also be coupled with such a term?

DEFINITION OF ARBITRATION AND JUDICIAL ROLE OF ARBITRATOR

Sutcliffe v Thackrah (1974) AC 727
In this case, the role of an architect was examined as to when it might be deemed to be that of an arbitrator.

From an analysis of the House of Lords' judgments, it appears that for an arbitration to take place:

1. There must be:
   (a) a specific dispute; or
   (b) present points of difference; or
   (c) defined differences that may arise in the future.
2. There must be a submission to an arbitrator or quasi-arbitrator.
3. The agreement must be binding.

As to the arbitrator, the burden of proof seems to be on him to show that he has been appointed to act as arbitrator by the terms of a contract or agreement between two parties to settle a dispute and is therefore acting in a judicial role and is immune from an action for negligence.

LIABILITY OR IMMUNITY OF ARBITRATOR TO SUIT FOR NEGLIGENCE

Arenson v Arenson [1975] 3 WLR 815

The series of judgments of the House of Lords in this case is considered to confirm that an arbitrator cannot be sued for negligence.

In fact most of the five judgments differ as to what constitutes an arbitration. Does there need to be a formulated dispute, whatever that may be, or a dispute, or is it sufficient that there should be a difference or opposing interests? Does the dispute need already to have arisen? Is it necessary that EVIDENCE or submissions are presented or can it be an arbitration where the arbitrator relied on his skill and judgement? Does the decision have to be binding?

Since their Lordships agree that immunity will be available only when an arbitration exists and there is no unity as to what is an arbitration then there is no certainty that immunity for arbitrators is available. This concept has yet to be tested.

Very few actions have been brought claiming negligence on the part of the professional advisers during a rent review. It is possible that most such claims never reach the court for the reason that dirty linen is best washed in private. This is also apparently true of solicitors, from whose interesting drafting most of these cases arise.

NEGLIGENCE OF EXPERT ADVISER IN ARBITRATION

Thomas Miller & Co v Richard Saunders & Partners [1989] 1 EGLR 267

There were two arbitrations relevant to the same premises. The arbitrator for both awards was the same. The rent fell to be reviewed both on the head lease and the sublease. The same agent acted in both arbitrations for the head lessee. The arbitration between lessor and head lessee was second.

There were corridors on each floor, most of which were required to be kept clear by the fire regulations for safety purposes. The parties in the arbitration concerning the subtenancy agreed that the corridors on the seventh and eighth floors should be excluded from the valuation. However, in the arbitration between the landlord and the head lessee the parties could not agree as to the seventh- and eighth-floor corridors. The agent for the head lessee had documentary evidence as to the requirement to retain these corridors unimpeded for fire purposes but failed to submit it to the arbitrator.

The arbitrator accepted the landlord's evidence on the corridors on face value and rejected the claims of the head lessee. As a result, the tenants had to pay a greater proportional rent than their subtenants. The lessees sought to sue their agent for negligence. They failed. Although in the award the arbitrator had separately stated that he rejected the evidence of the head lessee's agent with regard to the corridors, the award did not make it clear, in the court's eyes, as to the basis of the rejection. In fact, the court indirectly laid the blame upon the arbitrator for being inconsistent.

It is submitted that he was not inconsistent. In the arbitration relevant to the sublease the parties agreed the applicability of the corridors, whereas in the head lease arbitration the arbitrator was required to judge between the evidence of the parties. This he did. The fact that a representative failed to put in evidence which was apparently incontrovertible is surely the crux. As a result the arbitrator included the corridors in the valuation. The causation was as direct as need be in this case.

POSSIBLE NEGLIGENCE IN RENT REVIEW MATTERS AND A LATE NOTICE

Henniker-Major and Others v Daniel Smith (a firm) and Others [1991] 12 EG 58

A clear requirement in a review clause was for service of a rent notice by Lady Day 1980. The notice was not served until December 1980. The rent was eventually agreed. It was also agreed between the solicitors that by reason of the late service of the notice the new rent should only be payable from Lady Day 1981. There was a loss to the landlord therefore of the rent increase and interest. The landlord sued the surveyors, and subsequently his solicitors, for negligence.

The Court of Appeal decided that the concession to the tenant that the rent should only be payable from Lady Day 1981 was wrong. Time was not of the essence for service of the rent notice. However, the tenant should not be penalised by the late service of the notice when there was a rising market. There was no indication of what should happen in the event of a failure to serve a notice in time. The arbitrator made the correct decision in determining that the new rental figure should be valued as at the original date for service. This, the Court of Appeal decided, prevented any inequity to the tenants.

The Court of Appeal concluded that the rent was payable from the date when the notice should have been served, this in spite of the lease stating in clear terms that the increase should be ascertained as at the date of the service of the rent notice. This does imply that it should be payable also from that date.

This may not be the first time that inconclusive drafting has caused confusion out of all proportion to the initial ineptitude.

The Arbitration Act 1996 governs the procedure relevant to appeals both before the publication of an award and after the award has been issued.

APPEALS FROM ARBITRATION ON PRELIMINARY POINT OF LAW

Babanaft International Co SA v Avant Petroleum Inc. [1982] 1 WLR 871
A strong Court of Appeal set out the criteria for an appeal on a preliminary POINT OF LAW under the Arbitration Act 1979 s. 2. These applied even where there was an agreed application by both parties. LEAVE to appeal to the Court of Appeal should only be granted where:

1. the matter to be decided is one of general public importance (not a non-standard clause); and
2. the decision will substantially affect the rights of the parties;
3. and there will be a substantial saving in costs to one or both of the parties.

Also:

1. The High Court judge has a discretion to refuse to grant leave to appeal.
2. There is no need for a joint application.
3. The only instance where leave should be granted is where the point of law decides the whole dispute between the parties.

THE APPLICATION OF *BABANAFT* UNDER THE 1996 ACT

It is of some relief to know that the criteria both for appeals on a preliminary point of law as well as from the award itself have not changed in any great extent. Therefore, the citation from this apparently unreported judgment will mean that the *Babanaft* guidance yet applies.

Pirelli Construction Company Ltd. v Hamworthy Engineering Limited (4 July 1997, Queen's Bench Division, unreported)
The parties agreed in writing that the court could determine questions as to the substantive jurisdiction of the arbitration tribunal.

A further issue which was not a point of jurisdiction and had not been the subject of an agreement between the parties was sought to be decided.

The court considered its powers under s. 45 of the Arbitration Act 1996. It found that, although the resolution of the issue would both produce substantial savings in costs as between the parties and the matter did substantially affect the rights of the parties, without the consent of both parties the court did not have the jurisdiction to rule upon the issue.

The parties had chosen arbitration and to intervene would undermine the agreement made between the parties.

The court noted that the decision of the third issue would not decide even one issue in dispute between the parties. Therefore the use of the s. 45 procedure was quite inappropriate in the circumstances.

The guiding principles adopted by the court rely upon *Babanaft*.

APPEAL UNDER S. 2 OF ARBITRATION ACT 1979 – SAME RESULT EXPECTED UNDER 1996 ACT

Ritz Hotel (London) Ltd. v Ritz Casino Ltd. [1989] 2 EGLR 135

Although this appeal was made under the 1979 Act it is believed that the same result will emerge for appeals under the 1996 Act.

The test for an appeal to the courts on a preliminary POINT OF LAW is apparently quite strict. For discretion of the court to be granted the point of law must be a matter of public importance and the result must procure a considerable saving in costs.

In this case there were various questions to be decided as to what EVIDENCE might be used in DETERMINATION of the reviewed value and related matters which included the meaning of the phrase 'in so far as possible'. Although there is no question that all this was of considerable importance to the parties, it is hard to see how a rent review clause drawn up with particular reference to a casino is likely to be of general public importance. The court permitted an appeal.

It was also said that the test of considerable savings in costs to the parties was met. Surely the cost of addressing the learned judge cannot have differed greatly from that before the arbitrator.

Chapman v Charlwood Alliance Properties Ltd. (1981) 260 EG 1041

This was an appeal on a preliminary POINT OF LAW. *Chapman* was decided six days after *Babanaft* (see above), which was not cited. Any contradictions should therefore be viewed in that light.

The exercise of discretion by a judge in favour of granting LEAVE to appeal on a preliminary point of law when the application is made by only one party to the arbitration will only be considered:

1. if the consent of the arbitrator is obtained;
2. where the DETERMINATION of the application produces substantial savings in costs to the parties (not apparently just one party); and
3. where the point of law in question is one for which leave to appeal is likely to have been given had leave been sought under the Arbitration Act 1979 s. 1(3)(b) (appeal from an award) and not under s. 2 (appeal on a preliminary point of law before the award is issued).

Appeal from Award by Arbitrator under Arbitration Act 1996

As can be seen from the following examples, the scope for appeal arising out of a POINT OF LAW in the award is now much greater than it was a few years ago. It is a subtle exercise to attempt to separate what is fact from what is law and also what is a matter of mixed fact and law. Thus, in the guise of points of law, the courts have been asked to inquire whether a PREMIUM should be taken into account, whether a discount should be applied in certain circumstances, whether certain types of restaurant in differing areas fall into a single generic category, and other esoteric legal matters.

CRITERIA FOR GRANTING LEAVE TO APPEAL FROM AWARD

Pioneer Shipping Ltd. v BTP Tioxide Ltd. (The Nema) [1982] AC 724
This was a shipping case. The facts are irrelevant other than to say that a question of frustration of the contract arose. The arbitrator made a ruling being a decision of law. There was an attempt to seek LEAVE to appeal to the court against that ruling.

A surprisingly united House of Lords permitted Lord Diplock to set out the guidelines for granting leave to appeal if '… the dispute involves some question of law.'

Standard clauses   Where standard clauses require interpretation then:

1. where all parties agree, an appeal should *automatically* be heard; or
2. leave should be given at the request of one party where:
   (a) the arbitrator has misdirected himself in POINT OF LAW; or
   (b) … the decision was such that no reasonable arbitrator could reach it and where the rights of one or more of the parties have been substantially affected (Arbitration Act 1979 s. 1 (4)).

Where clauses are 'one-off'   Leave to appeal should be granted where all parties agree an appeal should automatically be heard; or

Where, as in the instant case, a question of law involved is the construction of a 'one-off' clause the application of which to the particular facts of the case is an issue in the arbitration, leave should not normally be given unless it is apparent to the judge upon a mere perusal of the REASONED AWARD itself without the benefit of adversarial argument, that the meaning ascribed to the clause by the arbitrator is obviously wrong. But if on such perusal it appear to the judge that it is possible that argument might persuade him, despite first impression to the contrary, that the arbitrator might be right, he should not grant leave.

… rather less strict criteria are in my view appropriate where questions of construction of contracts in standard terms are concerned…. But leave should not be given even in such a case, unless the judge considered that a strong *prima facie* case had been made out that the arbitrator had been wrong in his construction; and when the events to which the standard clause fell to be applied in the particular arbitration were themselves 'one-off' events, stricter criteria should be applied on the same lines as those that I have suggested as appropriate to 'one-off' clauses.

Suggested Criteria for Appeal

Standard clause	All parties	Appeal automatic
	One party	Leave needed
		Criteria:
		1. rights of party substantially affected and
		2. either self-misdirection or decision is such that no reasonable arbitrator should have reached it
'One-off' clause	All parties	Appeal usually automatic
	One party	Leave needed, very difficult to obtain
		Criteria as in standard clause

However, as will be seen, the subsequent decision in *Warrington* (see p. 101) wanders somewhat from these criteria. At present there does not appear to have been an analysis of the phrase 'rights of one or more of the parties substantially affected', although Donaldson MR implied that £70,000 was sufficient to affect the right of a party in *Aden Refinery Co Ltd. v Ugland Management Co* [1986] 3 All ER 737.

The *CMA* judgment of which extracts are set out below examines whether *The Nema* (see above) guidelines still apply. They do and therefore the cases noted below still apply, with the additional aspect that follows. It is the question of whether leave to appeal might be granted where the decision of the arbitrator raises a serious doubt (as to its validity).

CMA CGM SA v Beteiligungs-Kommanditgesellschaft MS 'Northern Pioneer' Schiffahrtgesellschaft mbH & Co and Others [2003] 1 All ER (Comm) 204, [2003] 1 Ll R 212
The judgment of the Court of Appeal was delivered by Lord Philips of Worth Matravers MR, who reviewed the history of s. 69 of the Arbitration Act 1996 and when LEAVE to appeal may be granted.

He cited *The Nema*, in which 'the House of Lords gave guidance as to the circumstances in which permission to appeal to the High Court from the decision of an arbitrator should be given. In relation to the construction of a "one-off" clause, permission should not be given unless, in the opinion of the court, the arbitrator was *obviously wrong*.'

His Lordship stated that 'Section 69 of the 1996 Act has replaced *The Nema* guidelines with statutory criteria.' He said that 'The statutory criteria are clearly strongly influenced by *The Nema* guidelines. They do not, however, follow these entirely. We have concluded that they open the door a little more widely to the granting of permission to appeal than the crack that was left by Lord Diplock.'

His Lordship noted that 'Section 69(3)(b) is an addition to *The Nema* guidelines, resolving a difference of view between the Commercial Court and the Court of Appeal in *Petraco (Bermuda) Ltd. v Petromed International SA* [1988] 3 All ER 454, [1988] 1 WLR 896.'

His Lordship then identified the key area which is additional to *The Nema* guidelines:

> Before he could grant permission to appeal s. 69 required that the judge should find (1) that the decision of the arbitrators on the existence of an implied term was obviously wrong or that the point was one of general public importance and that the decision of the arbitrators was at least open to serious doubt and (2) that reversing the decision of the arbitrators on the point would substantially affect the rights of one or more of the parties. We turn to consider these criteria. This question raises, however, consideration of the extent to which the criteria in s. 69 of the 1996 Act have departed from the *Nema* guidelines and this is a matter upon which we believe it would be helpful to give guidance.

His Lordship pinpointed the key departure in this way: 'The criterion for granting permission to appeal in s. 69(3)(c)(ii) is that the question should be one of general public importance and that the decision of the arbitrators should be at least open to serious doubt.'

Thus, the conclusion drawn is that the test discussed above being the strong *prima facie* case has been replaced by a test where the decision is open to serious doubts. This analysis will provide opportunities for a plethora of litigation. For what purpose did the legislators lessen the wise criteria described in *The Nema* and *Antaios Compania Naviera SA v Salen Rederierna AB* [1985] AC 191? It is difficult to arrive at a satisfactory conclusion.

AWARD – OBVIOUSLY WRONG – 1996 ARBITRATION ACT APPEAL ON A POINT OF LAW

Bisichi Mining plc v Bass Holding Ltd. [2002] EWHC 375 (11 February 2002, unreported)

The rent review machinery was lacking. The dispute concerned the date as to when the review provision should operate. Rent reviews could take place every twenty-one years. The date was either 1960 or 1962. It appeared that the phrase 'the said term' had different meanings throughout the lease.

The appeal was allowed on the basis that the arbitrator was obviously wrong. The judge, noting that the defending counsel could not easily defend the reasoning process in the award, did not identify the principles upon which the obviously wrong finding was made.

It is hoped that the judiciary will appreciate that their guidance is essential and without clear and simple direction the same errors can occur.

APPEALS FROM THE REASONED AWARD OF AN ARBITRATOR

Warrington & Runcorn Development Corporation v Greggs plc (1986) 281 EG 1075
An application for LEAVE to appeal by the landlords of a shopping centre was made, seeking a decision as to whether the arbitrator was correct in law in disregarding the effect of a PREMIUM when determining comparable rents. Leave was given on the following grounds:

1. The POINT OF LAW in question did affect the rights of 'a party' because resolution of this point could affect eleven other reviews in the centre and because the same arbitrator was likely to be appointed in all eleven. Therefore the 'one-off' criterion was met.
2. Further criteria were identified, namely, whether the judge was satisfied that the arbitrator could be wrong in law in the form either that there was a strong *prima facie* case that the arbitrator was wrong in law, or that the arbitrator was obviously wrong.

In spite of the fact that none of the above applied to the interpretation of a rent review clause, nor would the precise decision affect future reviews, or 'give rise to issue ESTOPPEL', the High Court still granted leave to appeal. The learned judge relied upon an interpretation of *The Nema* decision (see p. 98) which implied that Lord Diplock (and his fellow members of the House of Lords) concluded that the arbitrations being considered were only those of an international character. This is strange as that appeal arose under the 1979 Arbitration Act, which was intended to give finality to arbitrations where England and Wales were the forum, and which surely encompasses both domestic and international arbitrations.

Another unusual reason given for avoiding *The Nema* guidelines was that it did not matter so much if decisions on rent reviews were delayed. Some landlords might be surprised at that comment.

Ipswich Borough Council v Fisons plc [1990] 04 EG 127
The facts of this case do not apply to rent reviews. However, the Court of Appeal, led by the Master of the Rolls, Lord Donaldson – a past president of the Chartered Institute of Arbitrators – decided, indirectly, to clarify the confusion that had arisen since the decisions in *Lucas Industries plc v Welsh Development Agency* (1986) 278 EG 878 and *Warrington* (see above). The *Lucas* decision gave grounds for believing that an appeal under the auspices of rent review was more freely available than in other categories of arbitration.

The judgment of Lord Donaldson attempted to clarify the situation. The test for deciding to grant LEAVE to appeal was not whether there was a real doubt about the decision of the arbitrator on a POINT OF LAW. That test was too simple. The test was whether or not there was a 'strong' *prima facie* case for considering that the arbitrator was wrong in law.

Although this is also a judgment of Lord Donaldson, the test adopted in *CMA* (see p. 99), 'open to serious doubt', appears not be consistent here. This should not be a concern as *CMA* stated that there was somewhat greater scope for challenge than before.

As to the meaning of 'strong' in considering whether to permit the appeal, if the arbitrator were a lawyer employed for his expertise then the 'strong' would have to be very strong. If, on the other hand, the arbitrator were a non-lawyer whose expertise lay towards the main issue in dispute, and the legal issue were peripheral to the dispute, then also the 'strong' would have to be very strong. But if the arbitrator were a non-lawyer required to decide a legal issue as the main issue then the 'strong' would be less strong.

Finally, Lord Donaldson considered that there must be sufficient duplication of clauses to permit the claim that the point at issue was one of general public interest. This would permit compliance with the general test for appeals from awards of arbitrators.

This decision attempts to give finality to the awards of arbitrators in all but a very few cases. These few must meet the test laid down in the Arbitration Act 1979 and the amended tests set out in the 1996 Act, and *The Nema* case (see p. 98) and the decision in *Antaios Compania Naviera SA v Salen Rederierna AB* [1985] AC 191. Both these decisions were approved in *CMA*.

However, it appears that a new hurdle has arisen. First, it will be necessary to look at the qualifications of the arbitrator. In rent reviews, the arbitrator will usually be either a lawyer or a surveyor. If it is a lawyer then, where a point of law is fundamental to the dispute, an appeal will be much more difficult than if the arbitrator is a non-lawyer. A landlord, seeking a final award where a point of law is fundamental, will wish to appoint a lawyer, whilst the tenant will want a surveyor. If they do not agree, an appointment by the President of the RICS is likely to produce a surveyor. In order to ensure finality, landlords may try to appoint surveyors/lawyers by nomination within the terms of the lease. This would make it very difficult for either party to appeal against the award on any grounds other than misconduct.

CRITERIA FOR APPEAL – STRONG *PRIMA FACIE* CASE

Prudential Assurance Co Ltd. v Trafalgar House Group Estates Ltd. [1991] 01 EG 103

This judgment addresses the issue of the criterion that the rights of the parties must be substantially affected. This is an essential criterion under the 1996 Act.

A two-stage arbitration took place. The chartered surveyor arbitrator decided the value in alternative amounts of the revised rent, the alternatives depending on the definition of a disregard. Leading counsel then decided that definition.

The landlord appealed under s. 1(3) of the Arbitration Act 1979. There was an insubstantial effect upon the rights of the parties or at least one of them.

However, the parties agreed that their rights could be affected by the answer in the matter of this very disregard. A criterion of a successful appeal is that the rights of one or both parties should be affected substantially by the incorrect arbitral decision. Here neither party was substantially affected.

The Court of Appeal decided to ignore an earlier decision when an amount of one-eighth of the total was decided not to be sufficiently substantial to affect the right of the parties. Here the difference was only one-seventeenth. The reason the court ignored the earlier decision was that the judgment was 'taken in the early days of the Arbitration Act [i.e. 1982]. Furthermore the decision was given on the afternoon of October 1st and may possibly need to be reconsidered one day.' This is quite an extraordinary comment by the Court of Appeal. Indeed, what does it say about the leading House of Lords case, *The Nema* (see p. 98), which sets out all the criteria for appeals under the Arbitration Act 1979, most of which have been adopted in the 1996 Act? That case was also decided in 1982.

In the present case, the Court of Appeal decided that in any event the judge in the first instance used the wrong test of the validity of the award. He asked himself whether there was any real doubt that the award was correct. Whereas the correct test was to examine whether there was a strong *prima facie* case that the award was wrong. On that basis, the award was not wrong. There was no strong *prima facie* case that the leading counsel had erred. Therefore the judge should not have granted LEAVE to appeal. The appeal by the landlord therefore failed.

CRITERIA FOR LEAVE TO APPEAL

Capital & Counties plc v Hawa [1991] 46 EG 163
This is an example of an appeal which, it is hoped, would succeed under the 1996 Act. There were irregularities, the key irregularity being the giving of EVIDENCE to oneself, as arbitrator, where the parties were not given the opportunity to address the arbitrator. That should always be a successful ground of appeal except in the most trivial circumstances.

There was an appeal from the award of the arbitrator. The court decided the strict test of *Ipswich Borough Council v Fisons* (see above) was to be applied. There had to be a strong *prima facie* case that the arbitrator was unable to make the decision he did on the evidence before him.

The first ground of attack was that the arbitrator ignored the identical premises next door. He had the rental evidence and was aware that a PREMIUM had been paid for the premises in February 1990. The arbitration took place before November 1990 and related to a rent review valuation of 24 June 1990.

The arbitrator stated: 'The premiums paid upon assignment are unreliable indications of open market rental values and the transaction of limited assistance to me.' Similarly, he ignored the fact that the rent passing upon assignment was £1,550 higher than his valuation of the adjacent premises. There were no reasonings as to how the arbitrator devalued the comparable.

The test for the judge was whether on that evidence the rental value determined was one that no reasonable arbitrator could have reached upon the evidence before him. The learned judge found that it was not such an unreasonable DETERMINATION.

It is to be regretted that the landlord did not seek further reasons from the arbitrator. This might have led to a different conclusion by the judge.

The second challenge arose as to the deduction for management and other costs to be taken from the income of furnished letting. The tenants failed to submit any percentages. There was no indication as to the landlord's contributions. The arbitrator gave evidence to himself as to the percentage level of deduction. He did not allow the parties the opportunity to comment. In spite of this, the learned judge failed to find that such a determination was one which no reasonable arbitrator could have made; it was therefore not wrong in law. The award was upheld. Surely this is somewhat fortunate.

LEAVE TO APPEAL AND ASSUMPTIONS

British Gas plc v Dollar Land Holdings plc [1992] 12 EG 141
This is s. 1 of the Arbitration Act 1979 – appeal by consent. The consensual aspect for the right to appeal is reinforced under the 1996 Act.

No LEAVE of the court was required in this instance. The questions for the court to decide were, what were the terms of the hypothetical lease and when did the lease commence?

The relevant clause stated: 'for the purposes of this clause, a full market rent means the yearly rent at which the demised premises might reasonably be expected then to be let in the open market with VACANT POSSESSION by a willing lessor by a lease in the same terms in all other respects as this present lease (other than the original rents hereby reserved but including the provision for review of rent herein contained).'

The tenant defined the term as thirty-five years from the review date. The landlord, however, contended it meant thirty-five years from the original term. If the tenant were correct, the arbitrator offered a discount of 10 per cent on the current rental value.

The judge held against the tenant for various reasons, the first being that the use of the words 'same terms' meant the thirty-five-year original term. The fact that the rental values passing were to be excluded was also indicative of an original term. The learned judge found support for the common-sense approach from the Court of Appeal decision in *Basingstoke* (see p. 70). There it was indicated that the rent should be valued on the terms of a lease for a period unexpired when the review falls due.

The decision supported the landlord in that the term of years was that set out as being the original term. That meant that the lease had an unexpired term of fourteen years with implications for reversion thereafter.

Permission to appeal to the Court of Appeal against the judge's decision on a POINT OF LAW was refused. However, by reason of the large sums of money involved (more than a million pounds over seven years) a certificate was issued by the judge that there was a special reason why an appeal should be considered. It is of some concern that judges seem to be taking upon themselves the ability to consider that challenges to earlier decisions should be allowed only when large sums of money are involved. This, it is submitted, is a matter to be regretted.

APPEALS ON A POINT OF LAW

Fine Fare Ltd. v Kenmore Investments Ltd. [1989] 1 EGLR 143
Although this case was decided under the 1979 Act, it is a good example of an application for LEAVE to appeal where the criteria set out in the 1996 Act would also not be met.

The tenant attempted to obtain permission to appeal on a POINT OF LAW arising out of the award.

The tenant occupied a superstore at Luton. At the review comparable evidence was submitted in written form. It was common ground that superstores are valued either on the fitted-out or on the shell basis. The fitted-out basis is usually valued at ten per cent more than the shell basis.

The challenge pertained to a finding of fact concerning the comparable evidence.

As to whether a question of law arose, the learned judge examined the timetable of events and EVIDENCE. The landlord and tenant's agents disagreed as to whether the valuation upon which the landlord had relied had been made on a fitted-out or a shell basis. There was a conflict of evidence. However, it was impossible to show mathematically how the comparable had been valued in the award. It was equally impossible to decide whether the arbitrator had given any weight to HEARSAY EVIDENCE contained in a letter from the landlord's agent.

The learned judge decided that the mere possibility that hearsay evidence had been wrongly admitted was insufficient to permit remission of the matter to the arbitrator. The tenant failed in his application.

The judge stated 'I would only add that this case has demonstrated how important it is that arbitrator and experts should expressly state in their awards whether they are valuing premises on a shell or a fitted-out basis.'

My Kinda Town Ltd. v Castlebrook Properties Ltd. (1986) 277 EG 1144
The premises in question was a restaurant. It was the type of restaurant known as a 'destination' restaurant, i.e. one for which there is little passing trade, but for which there is an individual attraction. It was very successful.

At the rent review the landlord submitted various comparables of 'destination' restaurants which were outside the general area of the demised premises. The arbitrators took these comparables into account.

The appeal was made on the grounds that 'no person [the arbitrator] acting judicially and properly instructed as to the relevant law could have come to' the DETERMINATION which was made. The reason given was that the comparables had been trading successfully for a period of time, whereas the restaurant which was the subject of the review had only been successfully trading for a short time. The premises were assumed to be valued on a VACANT POSSESSION basis. This implied that those factors which contributed to its current success would not be present.

The court decided that the arbitrators' acceptance that the comparables were also destination restaurants was a finding of fact and as such unchallengeable. The assumption that the premises were vacant and to let did not prevent their availability as such a use being taken into account.

APPEAL FROM REFUSAL OF HIGH COURT JUDGE TO EXERCISE DISCRETION IN FAVOUR OF GRANTING LEAVE TO APPEAL

Aden Refinery Co Ltd. v Ugland Management Co [1986] 3 All ER 737
An application had been made to a judge of the High Court for LEAVE to appeal against the decisions of an arbitrator under s. 1(3) (b) of the Arbitration Act 1979. The judge refused to exercise his discretion.

The Court of Appeal stated that the judge could only exercise his discretion in favour if, in his view, there were a *prima facie* case that the arbitrators were wrong in law. Here the judge did not consider there was a *prima facie* case even though the arbitrators themselves considered that there were differing views on the issue of law. Donaldson MR decided that an appeal against the judge's refusal to exercise his discretion in favour of the appellant was in fact barred by statute.

The *CMA* judgment (see p. 99) may cast doubt upon this strict interpretation. It could be that the slightly lower test makes the exercise of discretion possible.

APPEALS – MAY THE PARTIES LITIGATE AGAIN ON THE SAME ISSUE?

Arnold and Others v National Westminster Bank plc [1991] 30 EG 57
The House of Lords considered circumstances when there might be right of appeal against the refusal of a judge to grant a certificate under s. 1(7)(b) of the Arbitration Act 1979. A certificate is required if the matter is going to be heard by the Court of Appeal. The certificate indicates that the matter is encompassed within the requirements of ss. 1 and 2 of the Arbitration Act 1979 and the comments made thereafter in *The Nema* (see p. 98).

Here the decision of the judge was considered erroneous. Anticipating a series of reviews over a twenty-year period, he considered they would be highly unfavourable to the tenant. The landlord was going to gain excessively by the surplus rent. Therefore the judgment was for the tenant.

Could there be an appeal or was it an issue of ESTOPPEL? In other words, should there be a legal process to prevent constant relitigation of the same issue between the same parties which would otherwise be an abuse of process? The House of Lords said that it would be an abuse of the process not to permit relitigation, this is in spite of the fact that the courts in England and Wales have aimed with the utmost determination at upholding the concept of the finality of arbitration awards.

What was the manifestly unfair decision? It was that a disregard encompassing 'other than the rent hereby reserved' was defined as including the rent provisions themselves, not merely the rental figure determined at the previous review. This does appear a somewhat inequitable decision. It is queried, however, as to whether, if the sum involved had not been of such significance, a twenty per cent differential on a million pounds, the House of Lords would have been sufficiently interested to make such a judgment.

It appears from this judgment that when a party can apparently find a situation where the decision of the judge is so perverse as to cause the fact of non-relitigation to be an abuse of process then the court will allow an appeal. It is suggested that much ingenuity will be exerted in order to create or find such situations.

On occasion, as we have seen, an award may be made which does not give reasons for the decision or if reasons are given they are not as full as might enable the parties to identify easily the reasoning behind the arbitrator's decision. Suggestions are made as to when further reasons can be requested.

ARBITRATOR'S AWARD AND REASONS

Leeds Permanent Building Society v Latchmere Properties Ltd. [1989] 1 EGLR 140

The USER CLAUSE within the thirty-nine-year lease restricted occupancy to a building society, and the first floor was to be occupied as offices. There was a local tenant restriction placed upon any possible occupiers of the first floor who might be a subtenant of the lessees.

A first rent review took place in 1983. A REASONED AWARD was published in which the arbitrator took into account the effect of the restricted user clause, and the possible difficulty of obtaining permission to assign to a use which was not that of a building society, in order to reduce the market rent by twenty per cent.

At a subsequent arbitration, reasons were not given. The rental level chosen was, within a very small percentage, that of the landlord. It was obvious that a minimal percentage discount had been applied, if any. The arbitrator had the benefit of seeing the first award, yet had apparently ignored the findings of that award.

The tenant challenged this award. As there were no reasons, the challenge was impossible unless the arbitrator could be persuaded to give reasons in spite of the fact that he was *functus officio* (no longer acting as an arbitrator). A successful challenge could be made only if the court could be persuaded under s. 1(5) of the Arbitration Act 1979 that a reasoned award should and would have been sought. The tenants applied for the court's discretion to be exercised in their favour.

The learned judge examined an allegation by the tenants that the refusal of the arbitrator to allow their 'counter counter-submissions' prevented further comment upon the evidence. The learned judge referred to a direction which specifically permitted the parties to seek an oral hearing which would have given ample opportunity for the parties to challenge the evidence. They did not do so.

It was noted in the judgment that arbitrators should not be required to give reasons weeks or even months after the event when memories have dimmed. A special reason for exercising the court's discretion had previously been deemed to have been where one party and a member of the arbitral staff had a genuine misunderstanding as to whether reasons were to be forthcoming or not. Here there was no confusion – merely negligence on the part of the representative of the tenant. The application failed.

A REASONED AWARD – WHEN MAY FURTHER REASONS BE SOUGHT?

Trave Schiffsahrtsgesellschaft mbH & Co AG v Nimemia Maritime Corporation (1986) 15 CSW 138

An arbitrator gave reasons without being requested so to do. He refused to give further and better reasons for the award when asked to do so.

A REASONED AWARD may not have been requested but an arbitrator may have given some reasons for his award. In that instance the court may require the arbitrator to give further and better reasons, but the court's discretion should be exercised bearing in mind the policy that an arbitrator should only be obliged to give reasons if asked to do so before the publication of the award. This was emphasised by the Master of the Rolls who said: 'in the absence of such a request [to give reasons] an arbitrator should not be expected to give reasons for an award after the award has been published. This was burdensome, an arbitrator should not be asked to undertake this unless it was vitally necessary for the purposes of an appeal.'

FURTHER ADVICE FROM THE COURTS AS TO WHEN ADDITIONAL REASONS SHOULD BE REQUIRED FOR UNREQUESTED REASONED AWARD

Warde v Feedex International Inc. [1984] 1 Ll R 310

The arbitrator gave a REASONED AWARD without having been requested to do so. There were insufficient reasons for the parties to be able to decide whether there were grounds for appeal. There were ISSUES of law in dispute. The court decided that in the exercise of its discretion to require further reasons of an unsought reasoned award it needed to consider:

1. What prospect there was of LEAVE to appeal being granted to request further reasons.
2. Whether there was anything in the conduct of the appellant which led to an exercise of the discretion against the appellant.

These comments do not detract from the reasoning in *Trave* (see case above), while the *obiter* guidelines give further assistance:

1. If one party requested a reasoned award the arbitrator should make a reasoned award save in very exceptional cases.
2. If both parties ask that there should not be a reasoned award, the arbitrator should also respect their wishes; but he should also, if asked, provide reasons in a separate document not incorporated in or forming part of the award.
3. If one party asked that there should not be a reasoned award and the other said nothing, the arbitrator should not make a reasoned award. But if it was doubtful whether the other party was aware of his rights, the arbitrator should consider whether it would be right to ask him.
4. Where nothing was said by either party, the arbitrator should again consider whether it would be right to ask the parties what form of award they wanted. Where the parties were represented by sophisticated advocates, he would be justified in assuming they wanted an award that would be without reasons.

ABILITY OF ARBITRATOR TO CHANGE AWARD

Pittalis v Sherefettin (1986) 278 EG 153

In this case the county court judge gave judgment for the plaintiffs having heard evidence and speeches and retired for twenty minutes. The next day the court officials wrote to the parties' solicitors saying that the judge had changed his mind and proposed to find for the defendants. The plaintiffs appealed.

The Court of Appeal refused the appeal and decided that a judge both in the County Court and in the High Court can withdraw his judgment where the judge has decided he was wrong as soon as the original judgment was given. The judgment should not be passed or entered.

There appears to be no reason why this judgment, as wide as it is, could not also be applied to awards of arbitrators provided they act precipitately.

Sometimes an arbitrator will in the course of the arbitration break the unwritten rules of natural justice. Depending upon the grossness of breach the party relying upon such misconduct or fault can apply to the courts for the award to be sent back (remitted) to the arbitrator for renegotiation or set aside altogether. Whenever a possible breach occurs, that breach should be brought to the attention of the other party and to the arbitrator. The arbitrator may not have been aware that there is a problem either by an act of omission or commission. There are examples of arbitrators' failing to insist upon adherence to directions. The aggrieved party must complain in writing and insist upon compliance before any further action takes place.

REMISSION TO THE ARBITRATOR AND DELAY

Arnold and Others v National Westminster Bank plc [1993] 01 EG 94

This case has a long history. Various appeals have taken place. The initial decision of the judge at first instance was found to be wrong by the House of Lords.

The tenants in this action attempted to have the original award overturned on the ground of misconduct. The problem for the court was that the tenants had missed the twenty-one-day appeal period following an award by three and three-quarter years.

This was too long in any event, in spite of the tenant's success in other issues. What the tenant should have done, the judge concluded, was issue all the relevant motions and summonses at the appropriate time and not tried various remedies in turn. The judge, therefore, refused this application. There might have been success in the instance of discovery of fresh EVIDENCE. This was not such a case. 'There was no suggestion in the authorities of any special provisions for, or possibility of, remission on the discovery of a subsequent decision pertinent indicating an error of law.' This was the situation here.

The tenant's submission, that remission should be 'available whenever justice cannot otherwise be done', was rejected by the learned judge. This situation will be enforced even more rigorously under the 1996 Act.

APPLICATION TO SET ASIDE THE AWARD OF THE ARBITRATOR ON THE GROUNDS OF MISCONDUCT

Handley v Nationwide Anglia Building Society [1992] 29 EG 123

There was a challenge to the award of the arbitrator by the landlord. The award was published on 16 October 1991. It was not taken up until 5 November 1991. Summons were issued by 19 November and dates given appropriate thereto.

The first application was for an extension of time to appeal against the award. The period of twenty-one days normally runs from the date of publication of the award. Here the judge decided that the applicable date was when the award was taken up (i.e. paid for) as opposed to published. The subsequent grounds of appeal for SETTING ASIDE were various. One issue was that the amount of the deduction arising from a planning restriction was not argued. The landlord considered that there should be no reduction, the tenant pleaded for ten per cent. The award included a reduction of 2.5 per cent. No comments had been sought from the parties. The judge disagreed on the grounds that the actual amount was *de minimis*. Thus, on practical grounds such niceties should be ignored.

The next point related to certificates of value. All comparable EVIDENCE should be accompanied by such certificates. If this is not the case, the comparable should not be admissible in evidence. The comparable was omitted, because the certificate was missing. The judge concluded, with evident validity, that this was not a miscarriage of justice.

A further matter was that on one comparable the arbitrator made a reduction of 19.5 per cent, partly because of a return frontage, without inviting comment from the parties. Also the arbitrator gave evidence to himself, as he had done earlier, as to the effect of a residential area on trading. For that he allowed a discount of ten per cent. That figure was not made available to the parties for comment either. As the arbitrator had taken the parties by surprise there was no question that he had not misconducted himself. The award was set aside. It is slightly to be regretted that all areas where arbitrators give evidence to themselves are not treated equally as effectively and strongly by the courts.

PROFESSIONAL NEGLIGENCE AND SOLICITORS AND THE MEASURE OF DAMAGES

Corfield v D.S. Bosher & Co [1992] 04 EG 127
There was an oral hearing in a rent review arbitration. By the second day of the hearing the landlord was envisaging an appeal. The hearing ended. By 21 January 1986 the arbitrator had informed the solicitors for both sides that the award had been published and was ready to be taken up (i.e. collected once paid for).

The solicitor for the landlord sent a copy of the letter to the landlord asking him to send a remittance to the solicitors for onward payment to the arbitrator.

Instead the landlord wrote directly to the arbitrator enclosing the cheque. The award was received by the landlord directly on 4 February 1986. The solicitor to the landlord wrote on two occasions reminding the landlord of the need for action. Not until 11 February did the landlord tell the solicitors that he had already taken up the award.

The time for appeal is twenty-one days. Time expired on 12 February. But the solicitors wrote to their client putting him on notice that time was short when in fact it had expired. The court decided that there had been negligence on the part of the solicitors in spite of the conduct of the landlord which had arguably but not unreasonably interfered with the normal process of payment.

The judge had then to decide whether to allow the out of time appeal. This he did by reason of the mistake, coupled with the fact that the landlord had been quick to appeal once he had been put on notice of the problems in the award. The judge had also to consider whether the arbitrator had so misconducted himself as to fall within s. 23 of the Arbitration Act 1950.

The first accusation was that the arbitrator took pedestrian counts and made observations which went outside his agreed instructions, thereby creating evidence upon which the parties were not able to comment. The judge decided that whilst the accusation was not as clear as in other cases, there was a small chance that the commercial court would have found misconduct and set the award aside. A very small chance, it is submitted. There were comparables. These it was incumbent upon the arbitrator to inspect. He did so and he assessed the pedestrian flow and attributed value to it. Most arbitrators would do likewise.

The second area of challenge came about as a result of the arbitrator applying a ten per cent discount to reflect the risk of planning permission being refused. This discount was not mentioned in the submissions of either side. Nor did the arbitrator mention his own thoughts to the parties such that they could address him on the issue. He gave EVIDENCE to himself.

The learned judge considered that the arbitrator was right to make an allowance for risk. And further he felt the point almost unarguable. It is submitted here that the judge is using incorrect criteria. Arbitrators must not give evidence to themselves.

The measure of damages related to the likelihood of success for an appeal was then decided. The judge determined the possible rental level obtainable if the landlord had successfully argued for a reassessment of the value of the comparables. He considered that the maximum lost was £23,380 over a seven-year period. But he only had a one-third chance of obtaining that figure. Therefore, the judge divided the original sum by three and came up with the figure of £7,790. To that he added a third of the costs of the appeal, £785. From this total should be deducted the costs of the next arbitration, which were estimated at £6,500. This surely is a strange decision as these costs would possibly have been borne by the tenant, had the arbitrator decided that the landlord was right. In any event, the various calculations produced a final sum of damages of £2,075. Not a great deal in all the circumstances.

OMISSION OF UPWARD ONLY REVIEW CLAUSE – APPORTIONMENT OF DAMAGES

Theodore Goddard v Fletcher King Services [1997] 2 EGLR 131

The plaintiffs asked the court to determine what percentage the defendants should contribute towards an agreed amount of damages due to a landlord.

The instruction was that an upward only review clause should be included in the rent review clause. It was omitted. The clause was to the effect upward or downward. The correct clause was inserted initially. Then the computer operator pressed the delete button (presumably) and the phrase was lost. The plaintiffs accepted liability. However they argued that the defendants also had a responsibility as the relevant person asked to see the draft as it progressed and did so. He also missed the omission.

The court decided that the admission by the plaintiffs did not resolve the defendants of responsibility. The court noted the difficulty of detecting an item which has been accidentally deleted. The percentage of responsibility was placed at twenty per cent.

In former times cross-checking documents in draft form was given to pupils or articled clerks, now this is deemed a waste of time. It is not.

The percentage chosen can only be arbitrary and sought at the behest of the respective insurance companies, presumably. An unnecessary expense where a common-sense approach would have split the damages equally.

WHEN IS AN ORAL HEARING MANDATORY? – MISCONDUCT OR PROCEDURAL MISHAP?

Control Securities plc v Spencer [1989] 1 EGLR 136

During a rent review the arbitrator issued various directions. He reserved the right to hold an oral hearing. Also the parties were to accompany their submissions with their comparables. Where there was only hearsay, second-hand, knowledge of the comparable then there was to be supporting confirmatory EVIDENCE. The arbitrator undertook to forward a copy of submissions to the other party. Further he undertook to contact the parties after receipt of the counter-submissions to discuss whether to hold an oral hearing.

The surveyor for the tenant did not comply with the direction as to comparables. Both the arbitrator and the landlord's agent drew attention to the omission. As a result a letter from the tenant was sent together with the counter-submission. The earlier problem was not rectified. A copy of the tenant's letter was not sent to the landlord. Nor did the agent of the landlord ask for a copy.

However the key area appears to have been the failure of the arbitrator to enquire whether the parties wished to have an oral hearing. As the court noted, if the hearing had occurred then the landlord's agent could have drawn attention to the omissions in the tenant's evidence in cross-examination.

The court set aside the award. This appears harsh as very similar facts in *Shield* (see p. 115) merited a remission to the arbitrator. The key is yet again that arbitrators and experts must not allow deviation from their directions by the parties or themselves unless all concerned agree in writing thereto. It should be noted that the criteria set out in the Arbitration Act 1996 are different. However, a reasonable perception of bias appears to be the necessary test.

Henry Sotheran Ltd. v Norwich Union Life Insurance Society [1992] 31 EG 70
In the arbitration it had been ordered that submissions should be made in the form of WRITTEN REPRESENTATIONS. Subsequently the tenant changed his surveyor. This surveyor expressly stated, when asked, that no amendments to the ORDER FOR DIRECTIONS were sought by him. However, he became uneasy at the some of the EVIDENCE of the landlord and asked for an oral hearing. The landlord's surveyor opposed this request but accepted that an oral hearing was an option.

The arbitrator accepted the right of either party to apply for an oral hearing. He suggested an oral hearing might be linked to certain ISSUES. These issues appeared to relate to matters of fact and opinion but not law. It is without doubt that where no evidence of fact or fact and opinion need be brought, that an oral hearing is unnecessary.

Time passed. A meeting took place. As a result the arbitrator issued a letter. That letter accepted that an oral hearing could take place if applied for in certain circumstances. A few days later he sent another letter asking the parties to make up their minds as to an oral hearing. Various extensions of time for response to this letter were granted.

A month later the tenant's surveyor requested an oral hearing. The arbitrator responded in an unfortunate manner. He suggested that the tenant's surveyor was using delaying tactics. The determination of dates for a hearing was problematic. A further letter suggested that the arbitrator was too busy to consider the representations at that time.

Suddenly, and without more ado, an award was issued. The judge had no difficulty in finding that misconduct had occurred. He set aside the award. The parties had been unable to apply to the courts to restrain the arbitrator from proceeding as they had no idea that an award was imminent. There was also a hint that the judge would have liked to ask the arbitrator to pay the costs of this litigation. However, the final award was costs in the further arbitration.

ARBITRATOR MAKES AN EXPERT DETERMINATION IN AN ADJACENT BUILDING – HAS HE IMPARTIALITY?

Moore Stephens and Co v Local Authorities' Mutual Investment Trust and Another [1992] 04 EG 135

An interesting point at issue here which must occur with frequency: an arbitrator who was appointed had previously been an independent expert in an adjacent comparable and had made the DETERMINATION. Neither party initially considered that the comparable would be of interest. Both parties were apparently aware of the arbitrator's role in the earlier matter.

Later the arbitrator agreed to permit that property to be used as a comparable. Forthwith the tenant made an application to have the arbitrator removed on the ground of partiality. It was made clear by the court that the integrity of the arbitrator was never in question in this litigation.

The first point raised was that the rental value determined by the then-expert was too high. The now-arbitrator would find it difficult to free his mind of that EVIDENCE in dealing with the current arbitration. This the trial judge rejected at once. That injustice might be done was rejected with equal rapidity by the judge.

Finally, the tenant desired to call the arbitrator as a witness. A witness of fact would be asked how he arrived at the comparable determination that he had made. This would entail the arbitrator being an expert witness. This evidence should be provided by the tenant's experts. Any factual evidence should be provided by the appropriate witnesses. An interesting legally justified application was rightly refused.

MISCONDUCT BY A JUDGE REFUSING AN ADJOURNMENT IN DIFFICULT CIRCUMSTANCE

Miah v Bromley Park Garden Estates Ltd. [1992] 10 EG 91

There was an application for a new tenancy. The tenant appointed solicitors. The day of the hearing drew near. In spite of frequent attempts the solicitors for the tenant could not obtain instructions. They did not attend the hearing. The tenant did. He stated that he was ill. He also stated that he was unable to contact his solicitor. Also he could not conduct his own case. He asked for an adjournment. This was refused. Later in the day he asked for a further adjournment. This was refused.

The judge himself noticed in the afternoon that the tenant did not appear to understand what he said to him. The tenant had a report from a chartered surveyor. But he did not have the chartered surveyor in the court itself. The report supported the contention of the tenant as to valuation.

The Court of Appeal decided to order a re-hearing. There was no opportunity for the tenant to cross-examine the expert of the landlord. He was incapable of doing so. The judge himself confirmed that the tenant was unable to understand what was happening. Injustice must be seen not to occur.

However, the tenant was ordered to pay into court the amount of the difference in rent between the sum claimed by the landlord and the sum contended by the tenant. This order was made as condition of LEAVE to appeal being granted and for leave to have a re-trial being permitted.

MISCONDUCT

Zermalt Holdings SA v Nu-Life Upholstery Repairs Ltd. (1985) 275 EG 1134
An arbitration unblemished in its procedure resulted in the production of a REASONED AWARD. The arbitrator in his award mentioned two matters not argued before him. Although these matters may not have influenced the award, it was right, the court decided, that where such matters appeared relevant to the arbitrator and these matters were not argued before him, he must always put these matters clearly to the parties before he makes his award. It was a breach of natural justice to use expertise without giving the parties an opportunity to comment. The court held that there was misconduct. The award was set aside although the misconduct, in this instance, was not much greater than that in *Shield Properties* (see below).

WHEN MISCONDUCT BY AN ARBITRATOR CAUSES REMISSION AND WHEN SETTING ASIDE OF AWARD

Shield Properties & Investments Ltd. v Anglo-Overseas Transport Co Ltd. (no. 2) (1986) 279 EG 1088
Here there was arbitration and subsequently an appeal from the arbitrator's award which was based on the submission of WRITTEN REPRESENTATIONS. There was the usual requirement that without prejudice negotiations should be excluded from any written submissions, and a requirement that all comparables should be accompanied by supporting EVIDENCE.

In the landlord's submission there was mention of without prejudice discussions. The tenant protested in writing but did no more. Subsequently, the arbitrator received from the landlord a list of comparables without supporting evidence. There was also a separate letter to the arbitrator offering this supporting evidence should he require it. The arbitrator did not send a copy of that letter to the tenant.

Here the court decided there was partial misconduct, but not sufficient to cause the award to be set aside. There were three possible areas of misconduct.

1. Admitting the without prejudice (or inadmissible) evidence. The court decided it was not misconduct because the tenants should have sought:
    (i)   either a ruling as to whether the submission was privileged; or
    (ii)   to stop proceedings; or
    (iii) to change the arbitrator; or
    (iv)   a clear understanding that all relevant evidence which was without prejudice should be ignored.
2. In this instance the tenants, other than in their protest, acquiesced by their conduct.
3. The court, however, decided that the arbitrator accepted evidence in a form contrary to his initial request, i.e. that all comparables should be accompanied by supporting evidence.
4. Finally, the court accepted that the arbitrator erred in not sending a copy of the vital letter to the tenants.

The court decided that, as to these last two errors, the arbitrator *did not misconduct himself* but should have kept in mind the need to adhere to the rules of natural justice, especially when lawyers were not involved. Therefore, although by his conduct he acted in a manner which was not more than procedurally irregular, the court decided that for such a minor matter it could be remitted to the arbitrator. This was done. The arbitrator subsequently valued the reviewed rent at a lower figure than that which he had put upon it originally.

Other judges might take a different view of the actions of the arbitrator and cause the award to be set aside rather than remitted. It could well be that the cost of a fresh hearing was not forgotten by the judge. However, misconduct by an arbitrator is not normally treated so lightly.

Arbitrators, it is understood, generally issue to themselves and their staff a check-list. Where ORDERS FOR DIRECTIONS are issued as well, on the check-list is the instruction that all correspondence to the arbitrator is to be copied to the other side. It is possible that this type of misconduct may eventually lead the party against whom the misconduct is perpetrated to bring an action against the appointing authority.

MISCONDUCT – TAKING ACCOUNT OF EVIDENCE THAT BOTH PARTIES AGREED SHOULD BE IGNORED

Oriel House BV v Lex Services plc [1991] 39 EG 139

Both sides agreed that implications for a rent-free period should be ignored. However, the arbitrator took the period into account in coming to his DETERMINATION. The landlord appealed.

A further concern was that EVIDENCE had been given by an expert that the level of uplift appropriate to a particular comparable should be greater than ten per cent. This had not been challenged, yet the arbitrator decided that the level of uplift was only 8.23 per cent.

Once again the arbitrator gave evidence to himself. He did not give the parties an opportunity to make submissions to him on these ISSUES and he did not give reasons for his decisions on them. Although the parties could have sought further reasons from the arbitrator they did not do so.

The learned judge came to the surprising conclusion that the fact that the parties agreed that certain evidence was not material did not prevent the arbitrator from using it and relying on it. The judge also felt that the point concerning the percentage differential was of little consequence, although the differential yielded a sum of over £20,000 plus VAT.

There was no misconduct and therefore no remission. This is unfortunate and will only encourage arbitrators to move outside the evidence submitted. Possibly, contracts of employment of arbitrators should spell out in explicit terms that arbitrators will not give evidence to themselves, or if they do they will be at risk as to costs.

FURTHER MISCONDUCT

Quinlan v London Borough of Hammersmith & Fulham (26 February 1985, unreported)
Various areas of alleged misconduct arose concerning the arbitrator, and there was an appeal from the award.

There was an intended inspection which occurred. The date of the letter indicating when the inspection would take place was only three days before the inspection was due. The lessee was on holiday over this specific period in any event. The partial inspection did take place by the arbitrator without any representative of the tenant being present (hardly possible for such representative to be present as the tenant was unaware of the impending visit). The arbitrator did not finally issue his award until approximately thirteen months after the DETERMINATION of the rent being made by him. It was said that:

1. He did not send copies of the lessor's comparables or submissions relating to rental values to the lessee, he merely notified the lessee of their existence.
2. He did not make a full inspection of the property, although he could have easily inspected the public areas even without the tenant's presence.
3. He did not send a copy of the award to the tenant.

The definition of misconduct was said by the judge to 'connote a procedural irregularity which either is, or at least gives the appearance of, being unfair'.

The learned judge decided that the following areas were misconduct:

1. Failure to supply copies of the documents of one party to the other. Such omission might be acceptable when both parties were experienced in arbitration matters, and here the case of *M.V. Myron v Tradax Export SA* [1970] 1 QB 527, was distinguished. Such omission, however, was not acceptable

when one party was a layman, 'since such a party would not have any ink-ling of the normal arbitration procedure and is likely to be totally unsophis-ticated in that sphere'. *Shield Properties & Investments Ltd. v Anglo-Overseas Transport Co Ltd.* (1984) 273 EG 69, was approved on this basis.

2. Failure to inspect the basement. The arbitrator, in affidavit EVIDENCE, indi-cated that both the basement and the ground floor were important. There-fore the court decided that it was not fair only to inspect the ground floor.

3. The arbitrator should not have confined himself to written submissions only. The learned judge quoted from Mustill and Boyd, *The Law and Practice of Commercial Arbitration in England* (Butterworths, 1982 edn.):

> You look to see if there are any express or IMPLIED TERMS in the agreement itself which lay down the sort of degree of informality which is required. Then if they (the terms) don't help you look at the subject matter, the matter of the dispute, the identity of the tribunal, bearing in mind that some (and here I am paraphrasing) types of arbitration, in particular quality arbitrations about goods in the City, may well merit the utmost degree of informality.

The court held in this case that although there was no express or implied term as to the degree of informality, here

> full Court formality [was] not required, but I do not think the very sub-stantial degree of informality adopted by the arbitrator, namely, written submission only was appropriate to the particular proceedings. It is very well known that comparables often start off heated arguments about whether the streets are the same, whether the environment is the same, and so on, and the comparables frequently become the subject matter of exten-sive cross-examination at hearings where they are relevant. Quite apart from that, it seems to me that a party (and this applied to both parties) might reasonably wish to give oral evidence concerning the condition of the premises, bearing in mind that any inspection can only reveal the situa-tion at the very moment when the inspection is made ... in my judgment, the arbitrator should have offered both sides the opportunity for an oral hearing if they wanted it.

It must be correct that arbitrators should follow a consistent pattern of pro-cedure, i.e. copying documents, etc. It is equally interesting that it was suggested that the parties should be offered the opportunity of an oral hearing. The following case also confirms the value of an oral hearing.

Littlewood & Another v Edwards & Others (Trustees for Knight Frank & Rutley (a firm)) (30 July 1981, unreported)
This was an application to set aside or remit the award of the arbitrator on the ground of misconduct. There were premises held on a lease dated November 1973, with a three-year review pattern. Parties failed to agree on the rent

payable from September 1980. An arbitrator was appointed. The arbitrator chose the method of hearing to be by WRITTEN REPRESENTATIONS. He subsequently suggested (not ordered) a timetable. The parties agreed dates by which submissions were to be exchanged. Dates were also agreed by which counter-submissions were to be submitted to the arbitrator. There was no requirement for these counter-submissions to be exchanged.

Two points are of interest:

1. That the parties asked the arbitrator whether he would wish to choose an oral hearing or not;
2. The parties agreed the timetable without any requirement to exchange counter-submissions.

Submissions were exchanged. The arbitrator asked for further evidence of a point made in the submission of the lessees. The lessors asked the arbitrator's permission to examine and comment upon this evidence. The arbitrator refused, indicating that the information he sought from the lessees did not need any further comment. The lessors tried again and failed. The lessees sent a considerable amount of further information to the arbitrator. The information was not disclosed to the lessors and a request for such disclosure to be made was refused by the lessees.

The lessors sent their counter-submissions to the arbitrator. They attempted to comment on what they believed the lessees had said on the submission of additional information. This information concerned the difficulty the tenants had in subletting the premises (by reason of the restrictive clauses set out in the lease) and their efforts made to this effect. The lessors considered the point important, and wished to throw doubt on the efforts of the lessees in this matter. They were not permitted to do so.

The award was published. In the award the arbitrator wrongly stated that both the submissions and the counter-submissions were exchanged. The lessors appealed against the award on the ground of misconduct.

The specific area of misconduct was the breach of rules of natural justice by the arbitrator in his failure to permit the lessors the same rights of comment and reply as he granted to the lessees.

The second alleged area of misconduct by the arbitrator was as follows: the lessors referred to the Hillier Parker Rent Index as providing an indicator as to the likely percentage increase in the levels of rent relevant to the West End of London where the premises were. The lessees did not demur nor did they submit an alternative index. The arbitrator, however, relied upon an index produced by the RICS which referred to rents applicable in the City of London but not to the West End. The percentage changes were less than those in the Hillier Parker Index.

The court decided that there was misconduct. The reliance by the arbitrator upon the incorrect index (not even submitted to him) was not a minor error. Also, as both parties considered important the ability to sublet easily, and as the arbitrator had apparently noted the lessees' contention on this matter, to refuse the lessors the right of comment was misconduct.

The court had then to decide whether the areas of misconduct were so severe as to cause SETTING ASIDE the award or whether REMISSION would be sufficient. No suggestion of lack of integrity by the arbitrator was made.

The award was set aside. The learned judge relied upon an element in the judgment of Sir John Donaldson MR in *Modern Engineering (Bristol) Ltd. v C. Miskin and Son Ltd.* [1981] Ll R 135, where he concluded that the correct test, as to whether to set aside an award or not, was whether either of the parties had lost confidence in the ability of the arbitrator to come to a fair and just conclusion should the matter be referred to him. The judge considered that the lessors would have lost this confidence. Therefore there was no choice other than to set aside the award in spite of consideration of additional costs and delay to the parties.

An important lesson to be learnt here is that where a timetable is agreed, neither party should exclude the right to see the counter-submissions or any items of correspondence of the other party.

The lessees were fortunate that the arbitrator sought further information from them, for they had themselves voluntarily given up their right to see any counter-submissions. The arbitrator by seeking amplification of a point without allowing the other side to comment was unwise and deserved censure.

Another point of interest is that the parties permitted the arbitrator to choose whether the hearing should be decided by written representations or by submission of oral evidence. If nothing else, this case supports the view that an oral hearing is to be preferred to that of written representations. If an oral hearing had taken place the parties would have been able in cross-examination to remove any possible areas of misconduct during the hearing itself (other than, of course, the arbitrator's reliance upon an index not submitted to him in the first place). (See the judge's comment in *Quinlan*, see p. 117.)

STATUS OF AWARD WHEN REMITTED BY THE COURTS

Shield Properties & Investments Ltd. v Anglo-Overseas Transport Co Ltd. (no. 2) (1986) 279 EG 1088

An arbitrator was required to determine the new rental level. The award had been challenged on the ground that the arbitrator had misconducted the proceedings (see pp. 115–16).

The award was remitted to the arbitrator to require the landlords to submit confirmation of comparable EVIDENCE. When this was done the rental level was reduced.

As the remittance might (and in fact did) affect the level of the rent, the court held that the entire award was 'in the melting pot'. Therefore, it was only when the 'second' award was issued that the rental level was determined. The first award was nullified in its entirety.

The award has now been accepted by the parties and a question at a subsequent review may arise as to the effect of the decision in the original award as to matters of both law and fact.

Award of Arbitrator and Subsequent Reviews

It is generally considered that an award of an arbitrator, whether it decides mere facts or a combination of facts and law, is not binding at subsequent reviews. Therefore it could be that where idiosyncratic points of law arise, these matters could be the subject of a diverse arbitral decision on each occasion. However, it is considered that where both parties sensibly agree to attach the arbitrator's award and the submissions relevant thereto to the lease such an attachment should, even in the event of an assignment, persuade the parties that the matter has been previously argued in sufficient depth to make unnecessary fresh and further disputes as to the niceties of meaning on fact and points of law.

It could be argued that when an ISSUE is referred to arbitration, the parties are prohibited from disputing that point again. It is agreed that the point in dispute at that very rent review is issue barred. However, is it issue barred at subsequent reviews, with the same parties but a different arbitrator and a different rental level? In a fresh dispute, surely all matters in dispute can be argued afresh. Without doubt, where there is a fresh party then there cannot be issue ESTOPPEL.

APPENDIX 1: GLOSSARY OF TERMS

ADMISSIBILITY   Rule of evidence which decides whether all relevant evidence can be admitted (given) in evidence; inadmissible evidence can be 'without prejudice' correspondence (see text).

BREAK CLAUSE   Clause in a lease or licence which permits either party to bring the contractual agreement to an end upon the service of a notice.

CONTRA-INDICATIONS   Clear or express words that state the desire of the parties that time shall be critical (of the essence) to some or all steps in the rent review process.

DEEMING   A phrase which indicates that the rent quoted shall be (deemed to be) the actual rent payable if a certain procedure is not followed.

DETERMINATION   The actual decision as to the rental value of the property in question.

DISCOVERY   The production of a list of documents in the possession of one of the parties to a dispute.

EQUITY   A concept which attempts to ensure that all parties involved in a dispute are treated fairly, sometimes in spite of legal constraints. This may be referred to as acting *ex aequo et bono* (fairly and in the best interests of all parties and ignoring legal concepts).

ESTOPPEL   A person is prevented (estopped) from denying a certain situation exists, e.g. that he has granted a lease, that a dispute has already been heard before between identical parties on identical issues.

EVIDENCE
(a)   Directly comparable – seldom available except on some industrial hi-tech estates and some shopping centres.
(b)   Best indirectly comparable – units in similar position, of similar size, with similar type of tenant.
(c)   Secondary indirect comparable – any of the above without one element.
(d)   Tertiary indirect comparable – generically similar units but without any of the best elements.

EX PARTE   An application to a judge or arbitrator made by one party without the other party being present.

GOODWILL   The trading benefit which arises merely from the existence of the business and its continuing profitable presence.

HEARSAY EVIDENCE   Speech that has been reported to someone else who then attempts to give evidence of that report.

IMPLIED TERMS   Terms which will be implied into an agreement in any event, usually arising from statutes.

ISSUES   Points in dispute between the parties.

LEAVE (of court)   Permission, as may be required, when wishing to appeal from an award of an arbitrator.

LEGAL ASSESSOR   A legally qualified person who may be called in to sit with the arbitrator or expert to advise on points of law (q.v.).

NON-SPEAKING AWARD   Award without reasons (see SPEAKING AWARD).

NOTICE OF MOTION   A legally prepared document which is given to (served upon) the other parties. The document will contain details of the place, date, and time when the party serving the document will ask a court for assistance in some way against the interests of the other parties. Thus, the other parties can as a result of the notice choose to be present or not at the time in question.

ORDER FOR DIRECTIONS   Set of instructions given by either consent or order of the arbitrator or by the independent expert for the procedure relevant to the determination (q.v.). It may include a timetable for exchange of documents, comment as to security for costs, and a limit upon the number of expert witnesses either side may call.

POINT OF LAW   The legal interpretation of a word of phrase.

PRECEDENTS   Judicial decisions which may influence a legal issue to be decided.

PREMIUM   A sum of money paid in addition to rent in various circumstances: e.g. to obtain sought-after premises; to release a covenant.

PRESUMPTION OF REALITY   This requires the valuer to acknowledge that there are real landlords and tenants in a contractual relationship in the premises, that the premises exist and are in the actual condition as at present, but then to take into account the assumptions and disregards as required by the terms of the lease.

REASONED AWARD   See SPEAKING AWARD

RECTIFICATION   Legal process where a judge may change the wording of a document.

REFERENCE   A dispute which arises between parties to a contract may be referred to a third party for a decision. The term 'reference' not only applies to the act of referral but also to the decision given.

REMISSION OF AWARD   The sending back of an award to an arbitrator, where the court has decided that the arbitrator has only made an insignificant error.

SETTING ASIDE AN AWARD OR DETERMINATION   Where the error in the award or determination, or in its procedure, is so fundamental as to prevent the court sending it back to its originator.

SPEAKING AWARD   An award to which the third party's reasons are attached. That the reasons are present enables either or both parties to the arbitration or reference (q.v.) to attack the award where one or more of the reasons given are incorrect.

SUCCEEDED ON TERMS   Succeeded but accepting that various conditions are placed upon that success which may be beneficial to the loser.

USER CLAUSE   Clause in lease or licence which attempts to state for what the premises may be used.

VACANT POSSESSION   Where a property is vacant and to let.

WAIVER   A waiver occurs when a party is aware that it has an opportunity to take a certain action but continues with the proceedings thus allowing the opportunity to pass. The question will frequently arise as to whether the opportunity has passed irrevocably. Depending upon the specific facts, the answer will generally be yes.

WRITTEN REPRESENTATIONS   A common method of determination (q.v.) of a reviewed rent whereby there is no oral hearing, merely an exchange of written submissions of evidence and such written amplification as may be permitted.

APPENDIX 2: EXAMPLE OF A NOTICE AND A COUNTER-NOTICE

Notice   20 September 2003

Dear Sirs,

Re: Lease of premises at 42 High Street, High Town, Low County

Under Clause 34 of the lease dated 30 November 1996 between yourselves and our clients, Righteous Limited, our client has the right to seek a reviewed rent at every fourth year from the commencement of the lease. Notice of his intention to seek such a rent must be notified to you in writing.

Therefore by this notice the landlord, our client, requires a reviewed rent for the demised premises for the period commencing 25 March 2004 of £20,000 per annum.

Yours faithfully,

cc: The landlord, Righteous Limited
cc: Solicitors to the landlord (or managing agent if this notice is issued by the landlord's solicitors)

Counter-notice   27 September 2003

Dear Sir,

Re: Lease of premises at 42 High Street, High Town, Low County

Thank you for your letter of 20 September 2003.

Our clients have noted the figure of reviewed rent required by your client for the period commencing 25 March 2004. Our clients do not accept the figure stated as being the value applicable to the demised premises. Therefore our clients reject the stated rent of £20,000 per annum.

As a result of this rejection our clients also request the appointment of an arbitrator (or independent expert) to determine the value of the demised premises as required by the procedure set out under Clause 44 of the lease dated 30 November 1996.

Yours faithfully,

INDEX